Applications of Nernst Equation

By Malika Ammam, PhD

Copyright© 2017 Malika Ammam. All rights reserved.

Discount Offers

5% OFF of the book price for purchases of 1-5 books.

8% OFF of the book price for purchases of more than 5 books.

To receive the discount money, send your request through https://www.malika-ammam.com/ with your order details and PayPal account. Make sure that your order details (amazon or other sites) passed the 30 days return policy.

Thank you,

Introduction

As a teacher of physical chemistry, I noticed that students, even in advanced classes, have difficulties in understanding the basics of redox chemistry. In this Section 6, I attempted to discuss some applications of the Nernst equation in electrochemical systems, including energy (consumption or generation), corrosion, redox titration, as well as the study of solubility, precipitation, and complexation processes. To further clarify the discussed concepts, numerous questions and problems with detailed answers are provided. Most of these questions are formulated by students like you. I believe that this Section 6 would greatly help students with levels varying from high school to advanced university classes.

Abstract

Redox equilibria are previously explained by the Nernst equation (Section 5), which could be applied in many electrochemical systems to estimate quantities, such as potentials and concentrations of redox species in solution. This section 6 is devoted to applications of the Nernst equation in electrochemical systems, including energy (consumption or generation), corrosion, redox titration, as well as study of solubility, precipitation, and complexation reactions.

1. Electrolytic cells

Electrolytic cells are a category of electrochemical cells consuming energy in the form of electricity to produce electrochemical reactions[1]. From the thermodynamic viewpoint, the redox reactions occurring in electrolytic cells are not spontaneous since external energy is used to produce these reactions ($\Delta G>0$ or $\Delta E<0$). The Nernst equation is suitable for use in electrolytic cells to determine quantities, such as potential, concentration, number of the electrons transferred during the process, as well as the electric charge ($q = -\frac{\Delta G}{E} = nF$, where q is the charge transferred and E represents the drop in potential between the two electrodes of the cell). Typical examples of electrolytic cells are those designed for electrolysis and electroplating.

1.1. Electrolysis

In electrolysis, external electric currents generated by power sources (e.g., power supplies, batteries) are employed to force chemical reactions to occur[1-3]. The passage of the electric current through the cell pulls out electrons from chemical species at the anode then drives them through the external circuit to the cathode, where they reduce other chemical species. In the internal circuit, charged cations and anions in the solution move toward the electrodes of opposite charge to neutralize the excess accumulated change at each pole and maintain the electroneutrality of the overall cell.

A typical example of electrolysis is water decomposition upon the passage of an electric current of at least 1.23 V through an aqueous electrolyte. Electrolysis of pure water forms an O_2 gas at the anode and H_2 at the cathode. The evolved gases generate bubbles at both electrodes. The two half-reactions and the overall redox reaction of water electrolysis are:

Oxidation: $2H_2O_{(l)} \rightarrow 4H^+_{(aq)} + O_{2(g)} + 4e^-$

Reduction: $2H_2O_{(l)} + 2e^- \rightarrow H_{2(g)} + 2OH^-_{(aq)}$

Overall reaction: $2H_2O_{(l)} \rightarrow 2H_{2(g)} + O_{2(g)}$

The electrolysis process does only apply to water but other substances can be electrolyzed as well. For example, molten salts like NaCl at 801 °C subjected to an electric current in a polarized cell will split into chlorine gas (Cl_2) at one electrode and deposit sodium metal at the other electrode.

Oxidation: $Cl^- \rightarrow \frac{1}{2}Cl_{2(g)} + e^-$

Reduction: $Na^+ + e^- \rightarrow Na_{(molten)}$

Overall reaction: $Cl^- + Na^+ \rightarrow \frac{1}{2}Cl_{2(g)} + Na$

When NaCl salt is dissolved in water, the electrolysis should induce hydrogen gas (H_2) at the cathode and chlorine gas (Cl_2) at the anode.

Oxidation: $2Cl^- \rightarrow Cl_{2(g)} + 2e^-$

Reduction: $2H^+_{(aq)} + 2e^- \rightarrow H_{2(g)}$ or $2H_2O_{(aq)} + 2e^- \rightarrow H_{2(g)} + 2OH^-_{(aq)}$

In other words, the presence of several competing ions in solution should first reduce or oxidize those with higher affinities to electrons (lower energies or potentials). For example, Na^+ might be reduced to Na in aqueous solution containing NaCl salt but this should require elevated energy (or potential), and because others species (e.g., H^+, H_2O) are present and require lower energies to reduce, they will be the species subjected to reduction. By comparison, in the molten state, only Na^+ and Cl^- are present in the media, thus they will the only species subjected to reduction and oxidation despite requiring higher energies (or potentials).

During electrolysis, the change in mass and number of electrons can be quantified by means of the Faraday's law of electrolysis. Quantitatively, the change in mass of a substance subjected to an electric current passed through an electrolyte solution is proportional to the quantity of the electric charge passed through the circuit. Also, the same amount of electricity passed through the cell should always induce the same change in mass of given species. On the other hand, the reduction or oxidation of 1 mole of a substance involving the transfer of one electron during electrolysis always requires 96485 coulombs of electric charge, known as the Faraday number.

The Faraday's law of electrolysis is expressed by:

$q = it = \frac{mFz}{M}$, or $m = \left(\frac{qM}{Fz}\right)$

where m is the mass of the substance in grams, q is the total electrical charge passed through the substance, i is the electrical current in ampere (A), F is the Faraday constant (F = 96485 C mol^{-1}), M is the molar mass of the substance in (g mol^{-1}), and z represents the valence of the substance or number of the electrons transferred per ion.

1.2. Electroplating

Electroplating is similar to electrolysis but instead of using alkali metal salts to transport the charge in solution, metal salts (e.g., $CuSO_4$, $FeCl_2$, $AuCl_3$) are employed. The application of an electric field to these metal cations induces their reductions at the cathode to form metal deposits[1,4]. Therefore, electroplating is often employed to modify conducting metal surfaces with other metals for various purposes, ranging from protection against corrosion to decoration. For example, electroplating is used in jewelry to coat nonexpensive metals (e.g., Ni, Cu) with thin layers of gold (Au). During this process, gold-based salts are dissolved in the electrolyte, which then are reduced by applying an electric field to the cell to deposit thin layers of gold ($Au^{3+} + 3e^- \rightarrow Au^0$) on the surface of the cheap jewelry metal. Also, several currencies are produced using electroplating, where copper coins are coated with nickel ($Ni^{2+} + 2e^- \rightarrow Ni$).

Overall, the combination of the Nernst equation with the law of electrolysis should allow determining changes in electrochemical quantities occurring in electrolytic cells.

2. Galvanic/voltaic cells (batteries)

Compared to electrolytic cells, galvanic cells produce energy (electricity) instead of consuming it. Thus, redox reactions in galvanic cells occur spontaneously ($\Delta G < 0$ or $\Delta E > 0$). In galvanic cells, the Nernst equation is also useful for quantifying quantities like potential, concentration, number of the electrons transferred, and the electric charge. Galvanic cells are also electrochemical cells, basically composed of two electrodes immersed in electrolytes to form two-half cells separated by a porous membrane (or salt bridge) to ensure ion migration and continuous flow of charge[2,5-6]. A salt bridge is simply an inverted U-tube filled with a non-reacting electrolyte and plugged at both ends of the cell with cotton or glass wool allowing ions diffusion but preventing electrolytes in both half-cells from mixing. The simplest salt bridge might be formed by a filter paper soaked in KNO_3 electrolyte.

In galvanic cells, the cell voltage is defined as the potential difference between the two half-cells, also known as the electromotive force (*emf*). The electrical current is generated by the flow (or transfer) of electrons from where they are produced to where they are consumed. The cell performance depends on both the current and potential (or *emf*) produced by the cell. This is quantified by the cell power: $P \propto I \times emf$, where P is the power, I is current, *emf* is the cell voltage (or potential), and $\propto$ symbolizes the proportionality between the parameters. Note that high voltage does not necessarily induce elevated current but the cell power is proportional to both voltage and current.

Compared to electrolytic cells, the notion of the anode and cathode are reversed in voltaic cells. In galvanic cells, the cathode is positive (+) and the anode is negative (-) whereas in electrolytic cells the cathode is negative (-) and the anode is positive (+). This is often expressed by cell notation (or diagram): Anode│Anode solution││Cathode solution│Cathode, where the single vertical line represents the change in phase solid/liquid/gas and the double vertical line indicates the separation between the two half-cells using porous barrier or salt bridge.

Current commercialized galvanic cells include primary (or non-rechargeable) batteries that could deliver electricity until all chemicals are exhausted, secondary batteries (or rechargeable batteries) that could be recharged and discharged for a number of cycles, and tertiary batteries (or fuel cells) that require continuous flow of oxidant and reductant to convert into electricity. Below are few selected examples of galvanic cells.

2.1. Daniell cell

The Daniell cell belongs to the category of primary wet cells and considered as one of the earliest assembled electrochemical cells[7]. The cell consists of two wires (Cu and Zn) placed in two half-cells containing $CuSO_4$ and $ZnSO_4$ electrolytes, respectively. The difference in potential between the Cu and Zn based half-cells induces oxidation of Zn at the anode due to its lower potential and reduction of Cu^{2+} at the cathode due to its relatively higher potential compared to Zn based half-cell. To prevent charge accumulation at both electrodes and maintain electroneutrality in both compartments, a porous barrier is placed between the two half-cells to ensure the migration of Zn^{2+} to the Cu compartment and SO_4^{-2} to Zn half-cell. The reactions occurring in the Daniell cell are:

Oxidation: $Zn_{(s)} \rightarrow Zn^{2+}_{(aq)} + 2e^-$

Reduction: $Cu^{2+}_{(s)} + 2e^- \rightarrow Cu_{(s)}$

Overall reaction: $Zn_{(s)} + Cu^{2+}_{(s)} \rightarrow Zn^{2+}_{(aq)} + Cu_{(s)}$

The generated electrons from Zn oxidation flow through the external circuit (conducting wire) and when they reach the Cu half-cell, Cu^{2+} reduces into Cu to deposit on the electrode. Thus, Zn slowly dissolves at the anode into the solution and copper deposits on the cathode to form a spongy brown film. As copper continues to deposit on the electrode, the characteristic blue color of copper sulfates gradually vanishes as its concentrations declines in the solution. In general, Daniell cells containing electrolytes at concentrations of 1M often induce cell voltages (or *emf*) of about 1.1 V coupled with substantial currents due to the low internal resistances of the cells. The shorthand notation (or diagram) of the Daniell cell is: $Zn|Zn^{2+}$ $(xM)||Cu^{2+}(yM)|Cu$, where x and y are respectively the concentrations of Zn^{2+} and Cu^{2+} in the two half-cells.

A closer look reveals that some components of the Daniell cell could be replaced by others. For example, the copper electrode could be replaced by any metal with similar or higher potential. Combinations of other metals to form similar cells are also possible. These include replacing Zn by Ni to form Ni/Cu cell or Cu by nickel to form Zn/Ni cell. The most important feature of the Daniell cell is that both metals have different potential to generate a flow of electrons between the two electrodes. In the proposed scenario, Ni could lose 2 electrons to form Ni^{2+} (Ni → Ni^{2+} + 2e^-, E^o = +0.25 V) and Cu^{2+} collects the two electrons to deposit on the electrode (Cu^{2+} + 2e^- → Cu, E^o = 0.337 V). This cell generates an *emf* of 0.59 V, which is less than that of the original Daniell cell (1.1 V). On the other hand, since the zinc sulfate solution plays only a role in ion conduction, it can be replaced by any ionic solution (e.g., NaCl, KCl). The porous barrier could also be substituted by a salt bridge (glass U tube filled with an electrolyte solution).

For decades, the Daniell cell was used as stationary power source in many fields, including telegraphic offices and some home appliances (e.g., doorbells).

2.2. Hydrogen electrode based cell

Instead of two-half cells made of metal electrodes like Zn and Cu, hydrogen electrode could replace one of them. The simplest hydrogen electrode based galvanic cell is made of an anode compartment comprising a Zn rod immersed in $ZnSO_4$ solution connected to cathode compartment containing a Pt rod filled with H_2SO_4 solution. At the anode, Zn oxidizes to Zn^{2+}, and the generated electrons pass through the external circuit (conducting wire) to reduce protons from the sulfuric acid and produce hydrogen gas at the Pt cathode.

Oxidation: Zn → Zn^{2+} + 2e^-

Reduction: 2H^+ + 2e^- → H_2

Overall cell reaction: $Zn_{(s)}$ + 2$H^+_{(aq)}$ → $Zn^{2+}_{(aq)}$ + $H_{2(g)}$

This cell delivers an *emf* of 0.76 V. Note that replacement of Zn with another metal with higher potential (e.g., Cu) should drive the overall reaction in the opposite direction because

Cu^{2+}/Cu has higher potential than H^+/H_2.

Oxidation: $H_2 \rightarrow 2H^+ + 2e^-$

Reduction: $Cu^{+2} + 2e^- \rightarrow Cu$

Overall cell reaction: $H_2 + Cu^{+2} \rightarrow 2H^+ + Cu$

This cell generates a voltage of only 0.34 V. Also, the sum of both above overall reactions results in the original Daniell cell. This is understandable since hydrogen electrode is used as a reference system and thus could easily be canceled.

$Zn_{(s)} + 2H^+_{(aq)} \rightarrow Zn^{2+}_{(aq)} + H_{2(g)}$

$H_{2(g)} + Cu^{+2}_{(aq)} \rightarrow 2H^+_{(aq)} + Cu_{(s)}$

Daniell cell: $Zn_{(s)} + Cu^{+2}_{(aq)} \rightarrow Zn^{2+}_{(aq)} + Cu_{(s)}$

2.3. Weston Cell

The Weston Cell also uses metal electrodes immersed in electrolyte solutions to spontaneously generate electricity according to the reactions:

Oxidation: $Cd_{(s)} \rightarrow Cd^{2+}_{(aq)} + 2e^-$

Reduction: $Hg_2SO_{4(s)} + 2e^- \rightarrow 2Hg_{(l)} + SO_4^{2-}_{(aq)}$

Overall cell reaction: $Cd_{(s)} + Hg_2SO_{4(s)} \rightarrow 2Hg_{(l)} + Cd^{2+}_{(aq)} + SO_4^{2-}_{(aq)}$

The Weston cell delivers an *emf* of about 1 V, which is similar to that of the Daniell cell[8].

2.4. Solid, moist cells (or dry cells)

Liquid cells like the Daniell, Weston and hydrogen electrode cells are useful for stationary usage but not for mobile appliances. Dry cells utilizing solid or moist paste components are more practical for movable appliances. A dry cell is composed of metal electrodes immersed not in liquid solutions but rather in mixtures of solid or moist pastes. At the anode, metals like Zn will oxidize to form Zn^{+2} and generate electrons. At the cathode, a carbon rod fixed in a paste containing ($MnO_2 + NH_4Cl + H_2O$) reduces MnO_2 to Mn^{3+}.

Oxidation: $Zn \rightarrow Zn^{2+} + 2e^-$

Reduction: $2MnO_2 + 8NH_4^+ + 2e^- \rightarrow 2Mn^{3+} + 4H_2O + 8NH_3$

Overall cell reaction: $Zn_{(s)} + 2MnO_{2(s)} + 8NH_4^+ \rightarrow Zn^{2+} + 2Mn^{3+} + 4H_2O + 8NH_3$

Because solid or moist cells are well sealed, they become more useful for use in movable or portable appliances and equipment. One limitation of these cells concerns the accumulation of ammonia around the carbon rod, which eventually decreases the electrical conduction since the gas has insulating properties.

In many of the mentioned liquid and dry cells, the reducing and oxidizing agents are depleted with time during the generation of electricity, yielding batteries with short life expectancies. To extend their lifetimes, rechargeable batteries and fuel cells are assembled in a way to either recycle products into reactants or continuously feed the compartments with reactants to generate electricity in the long run.

2.5. Reversible lead storage batteries

The lead acid storage batteries are considered as the oldest known reversible or rechargeable cells used in automotive starting engines[9]. In the charged state, the anode is made of Pb plate and the cathode from PbO_2 immersed in concentrated H_2SO_4 as the electrolyte. The redox reactions at both electrodes are summarized as follows:

Oxidation: $Pb_{(s)} + SO_4^{2-} \rightarrow PbSO_4 + 2e^-$

Reduction: $PbO_{2(s)} + 4H^+ + SO_4^{2-} + 2e^- \rightarrow PbSO_{4(s)} + 2H_2O$

Overall cell reaction: $Pb_{(s)} + 2SO_4^{2-} + PbO_{2(s)} + 4H^+ \rightarrow PbSO_4 + PbSO_{4(s)} + 2H_2O$

This cell delivers an *emf* of about 2 volts in the fully charged state. Note that standard batteries used for vehicle starting engines contain 6 cages of this cell assembled in series to yield a total *emf* of 12 V. Other cells with 6V could also be formed by connecting 3 cages of 2 V in series. Due to their low cost, these batteries are still being used in the automotive industry despite newly competitive technologies. One of their advantages is that the products induced by the oxidation of $PbSO_4$ and reduction of $PbSO_4$ both remain attached to the electrodes as solids without contaminating the electrolyte. This facilitates their reconversion during the charging process by using an external voltage to reverse the reactions. The shorthand notation (or diagram) of this cell is: $Pb \mid H_2SO_{4(aq)} \mid PbO_2$. Note that the vertical double line is absent because no separator is used in this cell.

2.6. Fuel cells

Fuel cells require a continuous source of fuel (H_2, ethanol, methanol, among others) and oxygen to be converted electrochemically into other chemicals and generate electricity[9-11]. As a result, fuel cells could produce electricity in a continuous manner as long as both the fuel and oxygen are supplied to the cell. The redox reactions of a H_2 fuel cell are:

Oxidation: $2H_2 \rightarrow 4H^+ + 4e^-$

Reduction: $O_2 + 4H^+ + 4e^- \rightarrow 2H_2O$

Overall reaction: $2H_2 + O_2 \rightarrow 2H_2O$

Note that fuel cells deliver low voltages, thus several cells must be assembled in series to multiply the *emf* and yield substantial voltages.

The classification of fuel cells is performed according to nature of the used electrolyte and startup times. Examples of fuel cells include proton exchange membrane fuel cells (PEMFCs), phosphoric acid fuel cell (PAFC), solid acid fuel cell (SAFC), alkaline fuel cell (AFC), solid oxide fuel cells (SOFCs), and molten carbonate fuel cells (MCFCs). Sometimes, biological catalysts (enzymes or microorganisms) are immobilized on the electrodes to accelerate the conversion of the fuel into products and generate electricity. Accordingly, these cells are called biofuel cells. Note that fuel cells have different efficacies in terms of cell performance, lifetime, and self-discharge.

3. Redox titrations

Redox titrations refer to the neutralization of certain amounts of oxidants by reductants. In redox titrations, an analyte solution containing an oxidant or reductant is often neutralized by a reductant or oxidant known as the titrant [12]. During this process, a certain number of electrons (or equivalent) is transferred. Note that an equivalent refers to the amount of a substance producing a change of one-mole oxidation number. At the neutralization point, the number of equivalents of the oxidant becomes equal to those of the reductant. Note that normality of the resulting solution is defined as the number of equivalent per liter of the solution.

The Nernst equation could also be used in redox titrations to determine parameters, such as redox potentials, concentrations, and the number of electrons exchanged during the titration process. Various titrants exist, including those based on salts of iodine I_2, bromine Br_2 and cerium (IV), as well as potassium salts of permanganates or dichromate. The addition of these titrants to solutions containing oxidants or reductants results in spontaneous redox reactions ($\Delta G < 0$) that require no external energy to occur. From the practical viewpoint, redox titrations are used in redox indicators and potentiometry, such as pH meters.

4. Solubility, precipitation, and complexation reactions

Sometimes, redox reactions involve soluble or slightly soluble species which renders the estimation of concentrations of dissolved species more complex[13]. The Nernst equation is very useful and could be utilized to estimate quantities like concentrations, ΔG, K_{eq}, E, and the number of the transferred electrons. For example, highly soluble salts (e.g., $FeCl_2$) will entirely dissociate to form ions (e.g., Fe^{2+} and Cl^-). Fe^{+2} could then be reduced or oxidized to other species (e.g., Fe or Fe^{3+}, respectively). By contrast, slightly soluble salts (e.g., AgCl) could not yield accurate measurements of the concentrations of dissolved ions (e.g., Ag^+). Thus, the Nernst equation under known conditions could be employed to estimate the concentrations of slightly soluble species.

For example, AgCl is slightly soluble in aqueous media and yields low concentrations of Ag^+, which could be reduced according to the reaction: $Ag^+ + 1e^- \rightarrow Ag$. The Nernst equation for this reaction could be written as: $E = E^o - \frac{RT}{nF} Ln \frac{(Ag)}{(Ag^+)} = E^o - \frac{RT}{nF} Ln \frac{1}{(Ag^+)}$ (because Ag is a solid (Ag) = 1). The measurement of the reaction potential will allow determining the concentration of Ag^+ in solution.

5. Corrosion

The Nernst equation can also be applied to corrosion processes to estimate concentrations or potentials. Corrosion of metals basically involves redox reactions, where metals oxidize in the presence of the reductant oxygen[14]. For example, the exposure of iron (Fe) to moisture and oxygen dissolves Fe into Fe^{2+} and generates electrons that are used to reduce oxygen. The process could be summarized as follows:

Oxidation: $Fe \rightarrow Fe^{2+} + 2e^-$

Reduction: $\frac{1}{2}O_{2(g)} + H_2O + 2e^- \rightarrow 2OH^-$

Overall reaction: $Fe + \frac{1}{2}O_{2(g)} + H_2O \rightarrow Fe^{2+} + 2OH^-$

The resulting Fe^{2+} and OH^- will combine together to form iron hydroxide precipitates or complexes depending on the conditions. The Nernst equation could be used to estimate parameters like reaction constants, concentrations of involved species, and potentials.

6. Combustion

The Nernst equation is also applicable to combustion reactions based on the oxidation of fuels by oxygen to yield carbon dioxide, water, and heat. Biochemical combustion uses enzymes as catalysts to accelerate the progression of the combustion reaction. For example, the combustion of carbohydrates and fats by oxygen in the human body induces carbon dioxide and energy used to maintain the healthy operation of vital functions, such regulation of body temperature, muscles contraction, and tissues build and repair. During this process, carbohydrates like glucose ($C_6H_{12}O_6$) and fats like tristearin ($2C_{57}H_{110}O_6$) convert as follows:

$C_6H_{12}O_{6\ (s)} + 6O_{2\ (g)} \rightarrow 6CO_{2\ (g)} + 6H_2O_{(l)}$

$2C_{57}H_{110}O_{6\ (s)} + 163O_{2\ (g)} \rightarrow 114CO_{2\ (g)} + 110H_2O_{(l)}$

Without catalysts, these combustion redox reactions are very slow and tend to be kinetically inhibited, thus rarely reaching equilibrium. Enzymes catalysts like glucose oxidase and lipase accelerate the kinetics of the above-mentioned reactions, respectively.

Summary

The Nernst equation is useful in estimating quantities, such as potential, concentration of redox species in solution, and the number of electrons exchanged in various redox equilibria. These include electrolytic cells, voltaic cells, redox titrations, and various reactions based on solubility, precipitation, complexation, corrosion, and combustion. Electrolytic cells (electrolysis and electroplating) require external energy (electricity) to produce redox reactions. Therefore, the reactions are not spontaneous ($\Delta G>0$ or $\Delta E<0$). By contrast, in voltaic cells, electrochemical reactions occur spontaneously to convert chemicals into electricity ($\Delta G<0$ or $\Delta E>0$). There are several types of voltaic cells, mainly classified into: i) primary (or non-rechargeable) batteries that could deliver electricity until all chemicals are exhausted, ii) secondary (or rechargeable) batteries which could be recharged and discharged for numerous cycles, and iii) tertiary batteries (or fuel cells) requiring a continuous flow of oxidants and reductants to convert into electricity. These batteries deliver variable voltages, currents, and have different performances. Some of them use liquid electrolytes (wet cells) and others solid or moist electrolytes (dry cells) more suitable for portable appliances, such as laptops and electric vehicles. Redox titrations allow neutralizing redox species with known concentrations with other species of unknown concentrations in an effort to identify the species and determine their concentrations. Redox titrations have suitable electron affinities to occur spontaneously without external energetic intervention. Some reactions based on solubility, precipitation and complexation involve redox species with very low concentrations, difficult to determine analytically. The Nernst equation could be used under these circumstances to estimate the concentrations of these species. The same applies to corrosion processes involving metal dissolution when the metal is subjected to a moist environment in presence of oxygen as a strong reductant. Finally, thanks to catalysts, combustion reactions could reach equilibrium at faster rates. The Nernst equation could also be used for these reactions to determine quantities, such as concentration and potential, among others.

References

1. Wendt, H.; Kreyse, G. (1999), Electrochemical Engineering: Science and Technology in Chemical and Other Industries, Springer.
2. Atkins, P. (1997). Physical Chemistry, 6th edition, W. H. Freeman and Company, New York.
3. Vanýsek, P. (2007), Electrochemical Series, in Handbook of Chemistry and Physics, 88th edition, Chemical Rubber Company.
4. Todd, R. H.; Dell, K. A.; Leo A. (1994), Surface Coating, Manufacturing Processes Reference Guide, Industrial Press Inc.
5. Atkins, P.; de Paula, J. (2006). Physical Chemistry, (8th edition). Oxford University Press.
6. Crompton, T. R. (2000), Battery Reference Book, 3rd edition.
7. Keithley, J. F. (1999), Daniell Cell, John Wiley and Sons, pp. 49-51.
8. Robert B (2005), Northrop Introduction to Instrumentation and Measurements, 2nd ed., CRC Press.
9. Linden, D.; Reddy, T. B. (2002), Handbook Of Batteries, 3rd edition, McGraw-Hill, New York.
10. Vielstich, W., et al., (2009), Handbook of Fuel Cells: Advances in Electrocatalysis, Materials, Diagnostics and Durability, Hoboken: John Wiley and Sons.
11. Behling, N. H. (2012). Fuel Cells: Current Technology Challenges and Future Research Needs (1st edition). Elsevier Academic Press.
12. Yong Zhou, Y. (2013), Redox Indicators: Characteristics and Applications, Elsevier.
13. Soustelle, M. (2016), Ionic and Electrochemical Equilibria, Wley.
14. Bardal, E. (2004), Corrosion and Protection, springer.
15. McAllister, S.; Jyh-Yuan, C.; Fernandez-Pello, A. C. (2011), Fundamentals of Combustion Processes, Springer Science & Business Media.

Section 6

Practical Questions and Problems with Solutions

A set of practical questions and problems with detailed solutions are provided to better understand the discussed concepts. The questions and problems range from simple to complex.

Q1. i) Describe the main components of the Daniell Cell. ii) At which terminals the oxidation and reduction take place? iii) Estimate the *emf* of the Daniell Cell. The standard potentials are: E^o (Zn^{2+}/Zn) = -0.76 V vs. NHE and E^o (Cu^{+2}/Cu) = 0.34 V vs. NHE.

Ans1. i) The Daniell cell is composed of two pots: one containing a Zn rod immersed in a $ZnSO_4$ solution and the other a Cu rod immersed in $CuSO_4$ solution. The two half-cells are connected with conducting wire and salt bridge for electrons and ions conduction, respectively.

ii) The electrons flow from the Zn electrode where they are generated because of the low potential of Zn relative to Cu. At the Cu electrode, the electrons are consumed by reducing Cu^{2+} into Cu. The oxidation and reduction half-reactions are as follows:

Oxidation: $Zn_{(s)} \rightarrow Zn^{2+} + 2e^-$, $E^o = +0.76$ V

Reduction: $Cu^{2+} + 2e^- \rightarrow Cu_{(s)}$, $E^o = +0.34$ V

Overall reaction: $Zn_{(s)} + Cu^{2+} \rightarrow Zn^{2+} + Cu_{(s)}$

Note that the Zn-based potential is reversed since the reaction is written in the oxidation form. The potential of the overall reaction (or *emf*) is simply the sum of potentials of both half-reactions written in their current states: $E^o_{total} = 0.76 + 0.34 = +1.10$ V

Q2. i) In your opinion, could redox chemistry be used to determine the solubility product of slightly soluble salts? If so, how? ii) AgCl is a slightly soluble salt, how could redox chemistry be used to determine the concentration of Ag^+ in solution? Calculate the potential of silver electrode immersed in (Ag^+) at 0.01M. iii) $Ag(CN)_2^-$ is a slightly soluble salt, what is relationship between the electrode potential and formation equilibrium constant of $Ag(CN)_2^-$?
The standard potential of E^0 (Ag^+/Ag) = 0.80 V vs. NHE.

Ans2. i) Yes, redox chemistry could be used to determine the solubility of slightly soluble salts using the Nernst equation: $E = E^o - \frac{RT}{nF} Ln\, Q = E^o - \frac{0.0592}{n} Log\, Q$, where Q is the reaction quotient linked to the activities (or concentrations) of the redox species.

ii) The reduction half-reaction involving Ag is: $Ag^+ + e^- \rightarrow Ag$

Therefore, $E = 0.8 - \frac{0.0592}{1} Log\, \frac{1}{(Ag^+)} = 0.8 - \frac{0.0592}{1} Log\, \frac{1}{(0.01)} = 0.68$ V (1)

iii) The formation reaction of $Ag(CN)_2^-$ complex can be written as:

$Ag^+ + 2CN^- \rightarrow Ag(CN)_2^-$

At equilibrium: $(Ag^+) = \frac{(Ag(CN)_2^-)}{K_{eq}(CN^-)^2}$ (2)

Combining Eqs. (1) and (2) gives: $E = 0.8 + 0.0592 \, Log \, (Ag^+) = 0.80 + 0.0592 \, Log \, \frac{[Ag(CN)_2^-]}{K_{eq}[CN^-]^2}$

Q3. Calculate the potential of the Daniell cell if the concentrations of (Cu^{2+}) and (Zn^{2+}) are respectively 0.01 M and 1.00 M. The standard potential of the Daniell cell is 1.1 V and the overall reaction is: $Zn + Cu^{2+} \rightarrow Zn^{2+} + Cu$

Ans3. The Nernst equation could be used to calculate the potential of a cell at conditions different from the standards. The Nernst equation corresponding to the overall reaction is:

$$E = E^o - \frac{RT}{nF} Ln \, Q = E^o - \frac{RT}{nF} Ln \, \frac{(Cu)(Zn^{2+})}{(Zn)(Cu^{2+})}$$

The number of the electrons transferred is n = 2, which could be determined by calculating the change in the oxidation numbers. The activity (or concentration) of solid metals, such as Cu and Zn metals is always 1.

Hence, $E = 1.1 - 0.0128 Ln \, \frac{1.00}{0.01} = 1.041$ V

Note that the potential of the cell at these concentrations is close to the standard value.

Q4. i) What is the potential of a cell composed of a Cu rod immersed in 1M $CuSO_4$ connected to a hydrogen electrode immersed in 1M HCl at 298 K (overall reaction occurs spontaneously)? ii) If the hydrogen electrode is replaced by silver electrode, what would be the cell potential? The standard potentials are: E^o (H^+/H_2) = 0 V vs. NHE, E^o (Cu^{+2}/Cu) = 0.34 V vs. NHE, and E^o (Ag^+/Ag) = 0.80 V vs. NHE.

Ans4. i) Since Cu has the highest potential, it will ensure the reduction half-reaction and hydrogen with lower potential will ensure the oxidation half-reaction.

Oxidation: $H_2 \rightarrow 2H^+ + 2e^-$, $E^o = 0$ V

Reduction: $Cu^{+2} + 2e^- \rightarrow Cu$, $E^o = 0.34$ V

Overall cell reaction: $H_2 + Cu^{+2} \rightarrow 2H^+ + Cu$

Note that the hydrogen-based potential is reversed since the reaction is written in the oxidation form. The cell voltage (or *emf*) is the sum of potentials of both half-reactions written in their current states: $E^o_{total} = emf = 0 + 0.34 = 0.34$ V

ii) Since the potential of Ag is higher, it will ensure the reduction and Cu will ensure the oxidation because of its lower potential. The two half-reactions involved in the process become:

Oxidation: $Cu \rightarrow Cu^{2+} + 2e^-$, $E^o = -0.34$ V

Reduction: ($Ag^+ + e^- \rightarrow Ag$) × 2, $E^o = +0.80$ V

Overall reaction: $2Ag^+ + Cu \rightarrow 2Ag + Cu^{2+}$

To eliminate the number of electrons in the overall reaction, the reduction reaction is multiplied by a factor of 2. The cell voltage (or *emf*) is the sum of potentials of both half-reactions written in their current states: $E^o_{total} = emf = +0.80 + (-0.34) = +0.46$ V

Q5. Consider a voltaic cell made by half-cell containing Mg rod immersed in Mg^{2+} solution connected to another half-cell containing Ag rod immersed in Ag^+ solution. i) Write down the reactions for a spontaneous flow of electrons, by specifying the oxidation, reduction and the overall reaction. ii) Provide a diagram for this cell. iii) Calculate the *emf* of the cell at the standard conditions. iv) Is this cell interesting from the application viewpoint? The standard potentials are: $E^o(Ag^+/Ag) = 0.80$ V vs. NHE and $E^o(Mg^{2+}/Mg) = -2.37$ V vs. NHE.

Ans5. i) Since the potential of Mg is very low compared to that of Ag, Mg will oxidize to generate electrons and Ag^+ will collect the electrons to reduce into Ag.

Oxidation: $Mg_{(s)} \rightarrow Mg^{2+}_{(aq)} + 2e^-$, $E^o = +2.37$ V

Reduction: $2Ag^+_{(aq)} + 2e^- \rightarrow 2Ag_{(s)}$, $E^o = 0.80$ V

Overall reaction: $Mg_{(s)} + 2Ag^+_{(aq)} \rightarrow Mg^{2+}_{(aq)} + 2Ag_{(s)}$

Note that the Mg-based potential is reversed because it is written in the reversed form (oxidation).

ii) The cell notation or diagram can be written as:

$$Mg_{(s)} \mid Mg^{2+}_{(aq)} \mid\mid 2Ag^+_{(aq)} \mid 2Ag_{(s)}$$

iii) The *emf* of cell is the sum of potentials of both half-reactions written in their current states: $E^o_{total} = emf = 2.37 + 0.8 = 3.17$ V

iv) The cell delivers quite a substantial potential that will fit with many applications, including automotive starting engines.

Q6. Provide a notation (or diagram) of a cell composed of copper rod immersed in Cu^{2+} solution connected to a zinc rod immersed in Zn^{2+} solution. The two half-cells are separated by a salt bridge and the overall reaction performs spontaneously. The standard potentials are: $E^o(Zn^{2+}/Zn) = -0.76$ V vs. NHE and $E^o(Cu^{+2}/Cu) = 0.34$ V vs. NHE.

Ans6. For spontaneous reaction, since Cu has a higher potential than Zn, Zn will oxidize and Cu^{2+} will reduce.

Oxidation: $Zn_{(s)} \rightarrow Zn^{2+}_{(aq)} + 2e^-$

Reduction: $Cu^{2+}_{(aq)} + 2e^- \rightarrow Cu_{(s)}$

Overall reaction: $Zn_{(s)} + Cu^{2+}_{(aq)} \rightarrow Zn^{2+}_{(aq)} + Cu_{(s)}$

The cell notation (or diagram) is: $Zn_{(s)} | Zn^{2+}_{(aq)} || Cu^{2+}_{(aq)} | Cu_{(s)}$

Q7. What is the cell notation (or diagram) of a cell composed of an aluminum rod immersed in Al^{3+} solution connected to a zinc rod immersed in Zn^{2+} solution. The two half-cells are separated by salt bridge and the overall reaction performs spontaneously. The standard potentials are: $E°(Zn^{2+}/Zn) = -0.76$ V vs. NHE and $E°(Al^{+3}/Al) = -1.66$ V vs. NHE.

Ans7. For spontaneous reaction, since Zn has a higher potential than Al, Al will oxidize and Zn^{2+} will reduce.

Oxidation: $(Al_{(s)} \rightarrow Al^{3+}_{(aq)} + 3e^-) \times 2$

Reduction: $(Zn^{2+}_{(aq)} + 2e^- \rightarrow Zn_{(s)}) \times 3$

Overall reaction: $2Al_{(s)} + 3Zn^{2+}_{(aq)} \rightarrow 2Al^{3+}_{(aq)} + 3Zn_{(s)}$

The cell diagram is: $2Al_{(s)} | 2Al^{3+}_{(aq)} || 3Zn^{2+}_{(aq)} | 3Zn_{(s)}$

To eliminate the number of electrons in the overall reaction, both the oxidation and reduction reactions are multiplied by factors that will give the same exact number of electron in each.

Q8. Provide notation (or diagram) for a cell composed of copper rod immersed in Cu^{2+} solution connected to Ag rod immersed in Ag^+ solution. The two half-cells are separated by salt bridge and the overall reaction is spontaneous. The standard potentials are: $E°(Ag^+/Ag) = 0.80$ V vs. NHE and $E°(Cu^{+2}/Cu) = 0.34$ V vs. NHE.

Ans8. For spontaneous reaction, since Ag has a higher potential than Cu, Cu will oxidize and Ag^+ will reduce.

Oxidation: $Cu_{(s)} \rightarrow Cu^{2+}_{(aq)} + 2e^-$, $E° = -0.34$ V

Reduction: $(Ag^+_{(aq)} + e^- \rightarrow Ag_{(s)}) \times 2$, $E° = 0.80$ V

Overall reaction: $Cu_{(s)} + 2Ag^+_{(aq)} \rightarrow Cu^{2+}_{(aq)} + 2Ag_{(s)}$

The cell diagram is: $Cu_{(s)} | Cu^{2+}_{(aq)} || 2Ag^+_{(aq)} | 2Ag_{(s)}$

Note that to eliminate the number of electrons in the overall reaction, the reduction reaction is multiplied by a factor of 2 to give the exact same number of electrons in each half-reaction.

Q9. Calculate the *emf* of $Mg_{(s)} | Mg^{2+}_{(aq)} || Cu^{2+}_{(aq)} | Cu_{(s)}$ voltaic cell at the standard conditions. The standard potentials are: $E°(Cu^{+2}/Cu) = 0.34$ V vs. NHE and $E°(Mg^{2+}/Mg) = -2.37$ V vs. NHE.

Ans9. To correctly calculate the *emf*, it is advised to first write down the two half-reactions with their respective potentials and the overall reaction. Next, sum the potentials to yield the overall cell potential.

The cell notation indicates that Mg is the oxidation terminal which can be confirmed by its lower potential and Cu is the reduction terminal justified by its higher potential.

Oxidation: $Mg_{(s)} \rightarrow Mg^{2+}_{(aq)} + 2e^-$, $E° = +2.37$ V

Reduction: $Cu^{2+}_{(aq)} + 2e^- \rightarrow Cu_{(s)}$, $E^0 = +0.34$ V

Overall reaction: $Mg_{(s)} + Cu^{2+}_{(aq)} \rightarrow Mg^{2+}_{(aq)} + Cu_{(s)}$

Note that the Mg-based potential is reversed because it is written in the reversed form (oxidation).

The cell potential (or *emf*) is the sum of potentials of both half-reactions written in their current states. *emf* = 2.37 + 0.34 = 2.71 V

This cell delivers substantial potential, which makes it interesting for energy devices.

Q10. Consider a cell composed of two half-cells: one consists of a Zn wire dipped in $Zn(NO_3)_2$ solution and the other of an inert electrode immersed in Fe^{3+}/ Fe^{2+} solution. Write down the two half-reactions and overall reaction. Calculate the cell *emf* at: 25 °C, (Zn^{2+}) = 0.22 M, (Fe^{2+}) = 0.42 M, and (Fe^{3+}) = 0.69 M. The standard potentials are: $E° (Zn^{2+}/Zn) = -0.76$ V vs. NHE and $E°$ $(Fe^{3+}/Fe^{2+}) = +0.77$ V vs. NHE.

Ans10. Since the potential of the Zn-based redox couple is lower, it will ensure the oxidation half-reaction and Fe-based redox couple will ensure the reduction half-reaction.

Oxidation: $Zn \rightarrow Zn^{2+} + 2e^-$, $E° = +0.763$ V

Reduction: $(Fe^{3+} + 1e^- = Fe^{2+}) \times 2$, $E° = +0.771$ V

Overall reaction: $2Fe^{3+} + Zn \rightarrow 2Fe^{2+} + Zn^{2+}$

The reduction reaction is multiplied by a factor of 2 to cancel the electrons in the overall reaction. The standard cell potential *emf°* is simply the sum of potentials of the two half-reactions written in their current states: *emf°* = 0.771 + 0.763 = 1.534 V

At conditions different from the standard, the Nernst equation is applicable:

$$emf = emf° - \frac{RT}{nF} Ln\, Q = emf° - \frac{0.059}{n} Log\, \frac{(Zn^{2+})(Fe^{2+})^2}{(Fe^{3+})^2} = 1.569\ V$$

(Zn) = 1 because it is solid.

Q11. Calculate the *emf* values of cells constructed by the following half-reactions. Which cells perform spontaneously?

$$Fe^{2+}_{(aq)} + 2e^- \rightarrow Fe_{(s)} \quad , \quad E = -0.44 \text{ V}$$
$$Al^{3+}_{(aq)} + 3e^- \rightarrow Al_{(s)} \quad , \quad E = -1.66 \text{ V}$$

Ans11. Since the direction of the electron flow is not specified, it is possible to construct two cells with these half-reactions. In the first cell, the oxidation occurs at the Al terminal and reduction at the Fe terminal. In the second cell, the reactions are reversed at both terminals.

First cell:

Oxidation: $2Al_{(s)} \rightarrow 2Al^{3+}_{(aq)} + 6e^- \quad , \quad E = +1.66 \text{ V}$

Reduction: $3Fe^{2+}_{(aq)} + 6e^- \rightarrow 3Fe_{(s)} \quad , \quad E = -0.44 \text{ V}$

Overall reaction: $2Al_{(s)} + 3Fe^{2+}_{(aq)} \rightarrow 2Al^{3+}_{(aq)} + 3Fe_{(s)}$

The cell potential is the sum of potentials of both half-reactions written in their current states:

emf = 1.66 + (-0.44) = 1.22 V

Second cell:

Oxidation: $3Fe_{(s)} \rightarrow 3Fe^{2+}_{(aq)} + 6e^- \quad , \quad E = +0.44 \text{ V}$

Reduction: $2Al^{3+}_{(aq)} + 6e^- \rightarrow 2Al_{(s)} \quad , \quad E = -1.66 \text{ V}$

Overall reaction: $2Al^{3+}_{(aq)} + 3Fe_{(s)} \rightarrow 2Al_{(s)} + 3Fe^{2+}_{(aq)}$

The cell potential is the sum of potentials of both half-reactions written in their current states:

emf = -1.66 + (0.44) = -1.22 V

In the first cell, the potential is positive, meaning that the Gibbs free energy ($\Delta G = -nFemf$) <0. Therefore, the overall reaction occurs spontaneously. In the second cell, the potential and free energy have reversed signs, meaning that the overall reaction is not spontaneous.

Q12. Given the two half-reactions with their standard potentials, calculate the *emf°* of a spontaneous cell.

$$A^{2+}_{(aq)} + 2e^- \rightarrow A_{(s)} \quad , \quad E° = -2.90 \text{ V}$$
$$B^{2+}_{(aq)} + 2e^- \rightarrow B_{(s)} \quad , \quad E° = +0.35 \text{ V}$$

Provide a diagram for this cell.

Ans12. If the reaction occurs spontaneously, the redox couple with the highest potential (B) will have more affinity to collect electrons, thus will ensure the reduction half-reaction. By contrast, the redox couple with the lowest potential (A) will have more affinity to lose electrons, hence will ensure the oxidation half-reaction.

Oxidation: $A_{(s)} \rightarrow A^{2+}_{(aq)} + 2e^-$, $E° = +2.90$ V

Reduction: $B^{2+}_{(aq)} + 2e^- \rightarrow B_{(s)}$, $E° = +0.35$ V

Overall reaction: $A_{(s)} + B^{2+}_{(aq)} \rightarrow A^{2+}_{(aq)} + B_{(s)}$

Note that the potential of the A-based reaction is reversed since it is written in the oxidation form.

The cell potential is simply the sum of potentials of the two half-reactions written in their current states. $emf = 2.9 + 0.35 = 3.25$ V

The potential of this cell is quite substantial, which makes it interesting for energy devices.

The cell diagram of the spontaneous cell is: $A_{(s)} | A^{2+}_{(aq)} || B^{2+}_{(aq)} | B_{(s)}$

Q13. Write down the two half-reactions and overall reaction of a cell performing spontaneously using the below redox reactions. Estimate the cell *emf* and provide a cell diagram.

$Zn^{2+}_{(aq)} + 2e^- \rightarrow Zn_{(s)}$, $E = -0.76$ V

$Cl_{2(g)} + 2e^- \rightarrow 2Cl^-_{(aq)}$, $E = +1.36$ V

Ans13. If the reaction occurs spontaneously, the redox couple with the highest potential (Cl) will have more affinity to collect electrons, hence will ensure the reduction half-reaction. By contrast, the redox couple with the lowest potential (Zn) will have more affinity to lose electrons, thus will ensure the oxidation half-reaction.

Oxidation: $Zn_{(s)} \rightarrow Zn^{2+}_{(aq)} + 2e^-$ $E = +0.76$ V

Reduction: $Cl_{2(g)} + 2e^- \rightarrow 2Cl^-_{(aq)}$ $E = +1.36$ V

Overall reaction: $Zn_{(s)} + Cl_{2(g)} \rightarrow Zn^{2+}_{(aq)} + 2Cl^-_{(aq)}$

Note that potential of the Zn-based reaction is reversed since it is written in the oxidation form.

The cell *emf* is simply the sum of potentials of the two half-reactions written in their current states: $emf = 0.76 + 1.36 = 2.12$ V

The cell diagram is: $Zn_{(s)} | Zn^{2+}_{(aq)} || 2Cl^-_{(aq)} | Cl_{2(g)}$

Q14. i) In few words define a dry cell. ii) For what applications are dry cells suitable? iii) Provide an example of configuration of dry cells by specifying the two-half and overall reactions.

Ans14. i) Dry cells are electrochemical cells which transform chemicals into energy through redox reactions. The particularity of dry cells is that the chemical components are either solids or in the form of pastes. ii) Dry cells are highly useful in portable appliances, such as flashlights, radios, toys, and other portable units.

iii) A typical example of dry cells consists of a Zn anode connected to carbon cathode surrounded by MnO_2. The electrolyte consists of moist paste of NH_4Cl and $ZnCl_2$

The two half-reactions and the overall reaction can be summarized as follows:

Oxidation: $Zn \rightarrow Zn^{2+} + 2e^-$

Reduction: $2MnO_2 + 8NH_4^+ + 2e^- \rightarrow 2Mn^{3+} + 4H_2O + 8NH_3$

Overall reaction: $Zn + 2MnO_2 + 8NH_4^+ \rightarrow Zn^{2+} + 2Mn^{3+} + 4H_2O + 8NH_3$

Q15. i) Briefly define corrosion of metals. ii) Provide a typical example of corrosion. iii) What factors could influence corrosion and what could be done to prevent it?

Ans15. i) Corrosion relies on the oxidation of metals in the presence of oxygen (O_2) and moisture (H_2O). This leads to deterioration of metals over time. ii) Rust is a typical example of corrosion, where iron (Fe) dissolves over time to form spongy deposited hydroxides/oxides.

iii) The factors influencing metal corrosion are: moisture (H_2O), oxygen (O_2), electrolyte (salts), and the presence of impurities in the metal. The abundance of oxygen, moisture, electrolytes (e.g., sea salt) and impurities in the metal should accelerate the corrosion process.

Hence, corrosion could be prevented by coating the surface with protective layers that limit the infiltration of O_2 and moisture. Examples include the application of hydrophobic substances (e.g., grease), painting, and/or modification of the surface with corrosion-resistant metal deposits.

Q16. Using redox reactions, briefly explain what is rusting of iron (Fe).

Ans16. Rusting of Fe relies on its corrosion in the presence of moisture and molecular oxygen (O_2). This leads to oxidation of Fe in the presence of O_2, a strong oxidant able of removing electrons from the Fe outer subshell. The reactions could be summarized as follows:

Oxidation: $Fe \rightarrow Fe^{2+} + 2e^-$

Reduction: $\frac{1}{2}O_2 + H_2O + 2e^- \rightarrow 2OH^-$

Depending on the conditions, the reaction often forms Fe hydroxides/oxides flacks deposited on the Fe surface.

Q17. i) Justify that corrosion of metals is a redox process. ii) What are the required conditions to initiate corrosion? iii) What are the factors which could accelerate the corrosion rate?

Ans17. i) Corrosion of metals is a redox process because it involves the transfer of electrons. O_2 as a strong oxidant is able to pull out electrons from the metal's outer shell. Thus, the metal is oxidized and O_2 is reduced during the process.

ii) The presence of air (O_2) and moisture (water) are necessary conditions for corrosion. The presence of air (O_2) in dry environment or wet conditions without oxygen will not induce corrosion.

iii) The presence of strong conducting electrolytes (e.g., sea salt) and higher temperatures could accelerate the corrosion rate.

Q18. i) Could aluminum (Al) be subject to corrosion? ii) How could iron be prevented from corrosion? iii) Both Al and Zn could be used to protect Fe against corrosion, why is Zn preferred over Al? iv) How could these metals be deposited on Fe?

Ans18. i) Al is often protected by a thin layer of Al oxide formed on its surface during the first stages of corrosion. This thin layer prevents the infiltration of O_2 and moisture into the material and limits further corrosion. ii) The modification of the iron surface could prevent its corrosion and prologue its lifetime. This could be performed by painting the surface, putting abundant grease on the surface, or electrochemically deposit a thin layer of non-corrosive metals on the surface of Fe. These modifications will prevent infiltration of O_2 and moisture into the inner layers of Fe and prevents (or decelerate) its corrosion.

iii) Zn is preferred over Al because it is cheap. iv) These metals are deposited on Fe using electrodeposition. Fe is immersed in metal salt solution (e.g., $ZnCl_2$, $AlCl_3$) then subjected to an applied voltage. This reduces Zn^{+2} or Al^{3+} from the solution to deposit as Zn or Al on the Fe surface and forms protecting coatings against corrosion.

Q19. Consider a cell composed of two half-cells: one contains a Cd rod immersed in Cd^{2+} solution at 0.03 M and the other a Pt rod immersed in Cl^- solution at 0.5M bubbled with Cl_2 gas at 1 atm. The overall cell is put in a chamber with controlled temperature of 10 °C. Identify the two half-reactions and estimate the cell *emf* and free energy. The standard potentials of the redox couples are: $E°$ (Cd^{+2}/Cd) = -0.4 V vs. NHE and $E°$ (Cl_2/Cl^-) = 1.36 V vs. NHE.

Ans19. Since the standard potential of Cl_2/Cl^- is superior to that of Cd^{+2}/Cd, the reduction will occur at the Cl_2/Cl^- electrode and oxidation at the Cd^{+2}/Cd terminal.

Oxidation: $Cd \rightarrow Cd^{+2} + 2e^-$, $E^o = +0.4$ V

Reduction: $Cl_2 + 2e^- \rightarrow 2Cl^-$, $E^o = 1.36$ V

Overall reaction: $Cd + Cl_2 \rightarrow Cd^{2+} + 2Cl^-$

Note that potential of the Cd-based reaction is reversed in sign because it is written in the oxidation form. At the standard conditions, the overall cell voltage (emf^o) is the sum of the standard potentials of the two half-reactions written in their current states:

$emf^o = 0.4 + 1.36 = 1.76$ V

The Nernst equation could be applied to calculate the voltage at conditions different from the standards: $emf = emf^o - \frac{RT}{nF} Ln\ Q$, where Q is the reaction quotient $\left(Q = \frac{(Cd^{2+})(Cl^-)^2}{(Cd)(Cl_2)}\right)$.

Since Cd is solid and Cl_2 is gas at 1 atm, both have concentrations (or activities) of 1. The number of the transferred electrons is n = 2.

This gives: $emf = emf^o - \frac{RT}{nF} Ln\ (Cd^{2+})(Cl^-)^2 = 1.76 - \frac{8.31 \times 283}{2 \times 96485} Ln\ (0.03)(0.5)^2 = 1.76 - 0.012 \times (-4.89) = 1.81$ V

The overall cell voltage under these conditions is slightly higher than the standard value due mainly to concentration effect.

The Gibbs free energy of the cell is:

$\Delta G = -nFemf = -2 \times 96485 \times 1.81 = -346.27$ kJ mol^{-1}

The free energy is negative, meaning that the overall cell reaction occurs spontaneously to generate energy.

Q20. Briefly define electroplating. In what major applications electroplating is used?

Ans20. In electroplating, a solution containing metal salts is subjected to an applied potential. The metal cations then reduce and deposit of the electrode to form thin films. For example, in electroplating of Zn, $ZnCl_2$ solution is subjected to an applied potential. Zn^{2+} then reduces to Zn ($Zn^{+2} + 2e^- \rightarrow Zn$) and thin layers of Zn metal are deposited on the electrode surface.

Electroplating is mainly used for decorative and protective purposes against corrosion.

Q21. Which of the following oxides/hydroxides are formed during corrosion of Fe: $Fe_2O_3 \cdot nH_2O$, Fe_2O_3, $Fe(OH)_3 \cdot nH_2O$, and/or $Fe(OH)_3$?

Ans21. Depending on the conditions (acidity, O_2, temperature), many of these oxides/hydroxides could form during the corrosion process of Fe.

Q22. Is the reaction between Zn and HCl redox? If so, identify the oxidant and reductant species?

Ans22. Electron transfer is involved in the reaction between Zn and HCl (metal attack by acid). Therefore, the reaction is redox. HCl is a strong acid and likely to be the oxidant pulling out electrons from Zn. The two half-reactions could be summarized as follows:

Oxidation: $Zn \rightarrow Zn^{2+} + 2e^-$

Reduction: $2H^+ + 2e^- \rightarrow H_2$

Overall reaction: $Zn + 2H^+ \rightarrow Zn^{2+} + H_2$

Since Cl^- is not involved in the reaction, it plays the role of a spectator ion.

The contact between Zn and the acid should liberate hydrogen gas.

Therefore, the oxidant is H^+ and the reductant is Zn.

Q23. Identify the false statement: i) rusting of iron is enhanced by moist air, ii) rusting of iron could be prevented by connecting it to an Mg rod, iii) rusting of iron is prevented by depositing a Zn coating, and/or iv) rusting of iron is prevented in the presence of NaCl.

Ans23. All the suggestions are correct except iv). NaCl is a highly soluble and conducting salt. As a result, its presence will rather accelerate the corrosion rate because it increases the transport of charge during the corrosion process.

Q24. i) Explain how the standard reduction potentials could help in determining the corrosion susceptibility of metals. ii) Explain why tin (Sn) is used in the manufacturing of canned containers of food preservation, such as canned sardines.

Ans24. i) The comparison between the standard reduction potentials should provide information regarding the metals that are more susceptible to oxidation. Metals with lower reduction potentials have more susceptibility to oxidation, hence corrosion.

ii) Since Sn (-0.14 V) has a higher potential than Fe (-0.44 V), it is used to modify Fe containers to increase their resistances against corrosion. Foods (e.g., sardines, vegetables, fruits, among others) could be stored for years without the cane possibly corrodes to cause food poising or change in food taste.

Q25. i) Explain why magnesium (Mg) is used as sacrificial metal electrodes against corrosion. ii) Propose a method to prevent ground pipes from corrosion.

Ans25. i) Mg could is used as sacrificial electrodes against corrosion because of its low potential (-2.71 V) compared to that of Fe (-0.44 V). Therefore, Mg could easily donate electrons when put in contact with O_2 instead of Fe.

ii) Ground pipes made of Fe can electrically be connected to Mg electrodes. Because of the low potential of Mg compared to that of Fe, it will corrode instead of Fe and protect the pipes from corrosion. The Mg electrodes have to be replaced by new ones as they become severely deteriorated by corrosion.

Q26. Consider a galvanic (voltaic) cell containing an unknown metal electrode X at the standard conditions: $X_{(s)} | X^{3+}(mol\ L^{-1}) || Pb^{2+}(mol\ L^{-1}) | Pb_{(s)}$

Assuming that the standard emf^o of the cell is 1.53 V, determine X. The standard potential E^o of $(Pb^{2+}/Pb) = -0.13$ V vs. NHE.

Ans26. The cell diagram suggests that oxidation occurs at the X terminal and reduction at the Pb terminal.

Oxidation: $(X \rightarrow X^{3+} + 3e^-) \times 2$

Reduction: $(Pb^{2+} + 2e^- \rightarrow Pb) \times 3$

Overall cell reaction: $2X + 3Pb^{2+} \rightarrow 2X^{3+} + 3Pb$

Note that the two half-reactions are multiplied by factors to eliminate the total number of electrons in the overall reaction.

The emf^o of the cell is the sum of potentials of both half-reactions written in their current states: $emf^o = E^o (X/X^{3+}) + E^o (Pb^{2+}/Pb)$, or $E (X/X^{3+}) = emf^o - E^o (Pb^{2+}/Pb) = 1.53 - (-0.13) = + 0.166$ V

Keep in mind that the obtained potential corresponds to the oxidation state, and to compare with standard reduction potentials found in thermodynamic tables, it has to be converted into the reduction state (-0.166 V). This potential corresponds to that of Al^{3+}/Al. Therefore, the other terminal is made by an Al rod immersed in Al^{3+} solution.

Q27. Briefly, what is a reversible cell? Provide a typical example of reversible cells.

Ans27. A reversible cell is an electrochemical cell which transforms chemicals into electricity through redox reactions. Contrary to first generation cells which have to be disposed upon discharge, reversible cells could be recharged after discharge by applying an external voltage to reverse the reactions. A typical example of reversible cells is lead batteries used in auto starting engines.

Q28. Explain the reason why dry cells could not work as reversible cells.

Ans28. In reversible cells, the products issued from the redox reactions could be reversed into their original forms by applying an external voltage to recharge the cell. To achieve this goal, the reaction products at the anode and cathode must stay attached to the electrodes to be reversed into reactants during the charging process. In dry cells, Zn^{2+} produced during oxidation of Zn diffuses away and ammonia gas issued at the cathode forms irreversible complexes. This makes it difficult to reverse the reactions to the initial reactants. As a result, once the cell is discharged, it cannot be recharged for another cycle.

Q29. i) What features voltaic cells require to recharge for other cycles? Justify your answer with an example. ii) What electrolytes are used in lead storage batteries?

Ans29. i) Rechargeable cells must have the reaction products stay in place at the electrodes for future reconversion into reactants during the charging cycles. If the products move away or form other complex components, they will be hard to reconvert into reactants. A typical example of rechargeable cells is lead storage batteries. The lead sulfates produced at the anode and cathode remain on the electrodes. Upon the application of an external voltage during the charging cycle, lead sulfate reconverts into the reactants lead and lead oxide at the electrodes. If the lead sulfate detaches itself from the electrodes and falls into the bottom of the cell during discharge, the reconversion will become impossible.

ii) The lead storage batteries use strong sulfuric acid solution (30%) as the electrolyte.

Q30. i) Write down the two half-reactions and overall reaction occurring in rechargeable lead batteries. ii) Estimate the *emf* of the cell. iii) How could this *emf* be converted into 12 V cell suitable for use in automotive starting engines? The standard potentials are: $E°$ (Pb^{2+}/Pb) = -0.13 V vs. NHE and $E°$ (PbO_2/Pb) = 1.455 V vs. NHE.

Ans30. i) Since the potential of the redox couple PbO_2/Pb (1.455 V) is higher than that of Pb^{2+}/Pb (-0.13 V), PbO_2 will reduce into Pb at the cathode and Pb will oxidize into Pb^{2+} at the anode according to the reactions:

Oxidation: $Pb_{(s)} + SO_4^{2-} \rightarrow PbSO_{4(s)} + 2e^-$, $E° = +0.13$ V

Reduction: $PbO_{2(s)} + 4H^+ + SO_4^{2-} + 2e^- \rightarrow PbSO_{4(s)} + 2H_2O$, $E° = 1.46$ V

Overall reaction: $Pb_{(s)} + 2SO_4^{2-} + PbO_{2(s)} + 4H^+ \rightarrow PbSO_{4(s)} + PbSO_{4(s)} + 2H_2O$

Note that the potential of the first half-reaction is reversed since it is written in the oxidation form. ii) The cell *emf* is the sum of potentials of the two half-reactions written in their current states: *emf* = 0.13 + 1.46 = 1.59 V

iii) This *emf* could be converted into 12 V cell by assembling several of the 1.59 V cells in series. This configuration will add their voltages to yield 12 V ($\frac{12}{1.59} = 7.54$). Therefore, assembling 8 of these cells should suffice to induce 12 V.

Q31. Consider a cell made a Zn rod immersed in Zn^{2+} solution as cathode connected to Pb rod immersed in Pb^{2+} solution as anode. i) Write down the two half-reactions and overall reaction. ii) Estimate the cell *emf* at the standard conditions. iii) Is the overall reaction spontaneous? The standard potentials are: E^o (Zn^{2+}/Zn) = -0.76 V vs. NHE and E^o (Pb^{2+}/Pb) = -0.126 V vs. NHE.

Ans31. i) Since the oxidation occurs at the Pb terminal and reduction at Zn terminal, the two half-reactions could be summarized as follows:

Oxidation: $Pb_{(s)} \rightarrow Pb^{2+}_{(aq)} + 2e^-$, E^o = +0.126 V

Reduction: $Zn^{2+}_{(aq)} + 2e^- \rightarrow Zn_{(s)}$, E^o = -0.76 V

Overall reaction: $Zn^{2+}_{(aq)} + Pb_{(s)} \rightarrow Zn_{(s)} + Pb^{2+}_{(aq)}$

ii) The cell *emf* is obtained by summing the potentials of both half-reactions written in their current states. Note that potential of the Pb-based reaction is reversed since it is written in the oxidation form. *emf* = 0.126 + (-0.76) = -0.634 V

The *emf* value is negative, meaning that the Gibbs free energy ΔG is positive (ΔG = - nF*emf*). The overall cell reaction is not spontaneous.

Q32. i) Briefly define redox titrations. ii) Estimate the molarity of H_2O_2 if 10 mL of H_2O_2 is required for neutralizing 20 mL of 0.1 M $KMnO_4$ solution. The standard potentials are: E^o (MnO_4^-/Mn^{2+}) = +1.58 V vs. NHE and E^o (O_2/H_2O_2) = +0.68 V vs. NHE.

Ans32. i) Redox titrations rely on neutralizing oxidants with reductants. Redox titrations are similar to acid/base neutralization but instead of proton exchange (acid/based), electrons are exchanged in redox titrations. ii) At neutralization point, the number of equivalents of the oxidant equals that of the reductant. In other words: $N_{H2O2} \times V_{H2O2} = N_{KMnO4} \times V_{KMnO4}$, where N represents normality and V the volume.

To determine the normality of each species, the two half-redox reactions and overall reaction should be written and balanced to view the stoichiometries.

The standard potentials indicate that the reduction occurs at the Mn terminal since it has a higher potential and oxidation at the H_2O_2 half-cell.

Oxidation: $(H_2O_{2(aq)} \rightarrow O_{2(g)} + 2H^+_{(aq)} + 2e^-) \times 5$

Reduction: $(MnO_4^-_{(aq)} + 8H^+_{(aq)} + 5e^- \rightarrow Mn^{2+}_{(aq)} + 4H_2O) \times 2$

Overall reaction: $5H_2O_{2(aq)} + 2MnO_4^-{}_{(aq)} + 6H^+{}_{(aq)} \rightarrow 5O_{2(aq)} + 2Mn^{2+}{}_{(aq)} + 8H_2O$

The overall reaction indicates that 5 moles of H_2O_2 require 2 moles of MnO_4^-.

Therefore: $\frac{1}{5}(M_{H2O2} \times V_{H2O2}) = \frac{1}{2}(M_{KMnO4} \times V_{KMnO4})$, where M is molarity.

$\frac{1}{2}(M_{H2O2} \times 10) = \frac{1}{5}(0.1 \times 20)$, or $M_{H2O2} = 0.16$ mol L^{-1}

Q33. i) Briefly define the equivalent weight of a redox substance. ii) Determine the equivalent weight of HCl in following redox reaction.

$Cl^- + 3H_2O \rightarrow ClO_3^- + 6H^+ + 6e^-$

iii) What are the equivalent weights of Na and Mg at their normal oxidation states?

Ans33. i) The equivalent weight of a redox substance represents its weight divided by the number of electrons involved in its oxidation or reduction process.

ii) The reaction indicates that the oxidation state of Cl changed from -1 to +5. Hence, 6 electrons are exchanged during the process. In this case, the equivalent weight is one-sixth ($\frac{1}{6}$) of the molecular weight of HCl or: $\frac{36.5}{6} = 6.08$ g

iii) At normal oxidation states, Na loses 1 electron and Mg can donate up to 2 electrons following the reactions:

Na $\rightarrow$ $Na^+ + e^-$

Mg $\rightarrow$ $Mg^{2+} + 2e^-$

The reactions indicate that the equivalent weight of Na equals to its atomic weight, and that of Mg is $\frac{1}{2}$ of its atomic weight.

Q34. Estimate the equivalent weight of $KMnO_4$ during its redox titration with H_2O_2 following the reaction:

$2MnO_4^- + 5H_2O_2 + 6H^+ \rightarrow 2Mn^{2+} + 5O_2 + 8H_2O$

Ans34. The overall reaction indicates that the oxidation number of Mn decreased from +7 in MnO_4^- to +2 in Mn^{2+}. This means that 5 electrons are exchanged during the process.

Equivalent weight = $\frac{\text{molar mass}}{\text{number of electrons gained or lost}} = \frac{158}{5} = 31.6\ g$

Q35. i) If a hydrogen electrode is connected to Zn electrode, what will be the oxidation and reduction half-reactions and the overall reaction? ii) Write down the two half-reactions and overall reaction if Zn is replaced by Cu. iii) Name this cell. iv) Calculate the cell voltage (or *emf*) in both cases. The standard potentials are: E^o (Zn^{2+}/Zn) = -0.76 V vs. NHE, E^o (H^+/H_2) = 0 V vs. NHE, and E^o (Cu^{+2}/Cu) = 0.34 V vs. NHE.

Ans35. i) Because Zn has a lower potential than hydrogen, Zn will oxidize and H^+ will reduce according to reactions:

Oxidation: Zn → Zn^{2+} + 2e^- , E^o = +0.76 V

Reduction: 2H^+ + 2e^- → H_2 , E^o = 0 V

Overall reaction: $Zn_{(s)}$ + 2$H^+_{(aq)}$ → $Zn^{2+}_{(aq)}$ + $H_{2(g)}$

Note that the Zn-based potential is reversed since the reaction is written in the oxidation form. The cell voltage (or *emf*) is the sum of potentials of both half-reactions written in their current states: E^o_{total} = 0.76 + 0 = 0.76 V

ii) Replacement of Zn by Cu reverses the flow of electrons since Cu has higher potential or more affinity to electrons than hydrogen. As a result, Cu will reduce and hydrogen will oxidize to supply the electrons.

Oxidation: H_2 → 2H^+ + 2e^- , E^o = 0 V

Reduction: Cu^{+2} + 2e^- → Cu , E^o = 0.34 V

Overall reaction: H_2 + Cu^{+2} → 2H^+ + Cu

iii) Cells involving hydrogen electrode are called hydrogen electrode cells. iv) Note that the hydrogen-based potential is reversed since the reaction is written in the oxidation form. The cell voltage (or *emf*) is the sum of potentials of both half-reactions written in their current states: E^o_{cell} = emf^o = 0 + 0.34 = 0.34 V

Q36. Are redox potentials useful in estimating solubility constants of slightly soluble salts?

Ans36. Yes, redox potentials are useful in estimating the solubility constants of slightly soluble salts because they are challenging to measure analytically due to the very low concentrations. This could be achieved by means of the Nernst equation.

Q37. Consider a cell made of two inert metal rods: one immersed in half-cell containing (Fe^{3+}/Fe^{2+}) and the other in second half-cell containing (Br_2/Br^-). At which electrodes will the oxidation and reduction occur if the overall cell reaction is spontaneous? Calculate ΔG of the overall reaction. The standard potentials are: E^o (Fe^{3+}/Fe^{2+}) = 0.77 V vs. NHE and E^o (Br_2/Br^-) = +1.066 V vs. NHE.

Ans37. Since both electrode materials are inert, they will not be involved in the redox reactions, except for transferring the electrons from one pole to the other.

Comparison between the potentials of the two half-reactions indicates that Br_2/Br^- will ensure the reduction half-reaction because of its superior potential.

Oxidation: $(Fe^{2+} \rightarrow Fe^{3+} + 1e^-) \times 2$, $E^o = -0.77$ V

Reduction: $Br_2 + 2e^- \rightarrow 2Br^-$, $E^o = +1.066$ V

Overall reaction: $Br_2 + 2Fe^{2+} \rightarrow 2Br^- + 2Fe^{3+}$

Note that the oxidation reaction is multiplied by a factor of 2 to eliminate the number of electrons in the overall reaction. However, the redox potential should not be multiplied because the number of electrons is already included in its calculation ($E = -\frac{\Delta G}{nF}$). On the other hand, calculation of ΔG should include the overall electrons exchanged in the overall reaction.

The *emf* of the cell is simply the sum of potentials of both reactions written in their current states:

emf = +1.066 - 0.77 = 0.296 V

Hence, ΔG = -nFE = - 2 × 96.5 × 0.296 = -57.128 kJ mol^{-1}

The negative sign of ΔG confirms the spontaneity of the reaction.

Q38. A cell is made by two inert conducting rods immersed in acidic solution. H_2 gas is bubbled at one rod and Cl_2 at the other under the standard conditions. Identify the oxidation and reduction poles. Calculate the cell *emf* and free energy. The standard potential of (Cl_2/Cl^-) is +1.36 V vs. NHE.

Ans38. Since the metal rods are inert, they will only transfer the electrons from one pole to the other and will not participate in the redox reactions. Consequently, the possible reactions at the anode and cathode will involve H_2 gas, Cl_2, and protons H^+ present in the solution. The potential of the redox couple (H^+/H_2) at the standard conditions is 0 V and that of (Cl_2/Cl^-) is +1.36 V. Hence, (Cl_2/Cl^-) should be the reduction pole because of its elevated potential and (H^+/H_2) the oxidation pole.

Oxidation: $H_2 \rightarrow 2H^+ + 2e^-$, $E^o = 0$ V

Reduction: $Cl_2 + 2e^- \rightarrow 2Cl^-$, $E^o = +1.36$ V

Overall reaction: $H_2 + Cl_2 \rightarrow 2H^+ + 2Cl^-$

The cell potential or *emf* is the sum of potentials of both half-reactions written in their current states: *emf* = 0 + 1.36 = 1.36 V

The ΔG of the cell is defined by: ΔG = -nF*emf* = -2 × 96.5 × 1.36 = -262.48 kJ mol^{-1}

The negative sign of the free energy means that the overall reaction is spontaneous in this direction.

Q39. Consider a cell composed of Fe and Pt rods immersed in acidic solution. The Pt rod is bubbled with hydrogen gas. Determine the two half-reactions and overall reaction if the process is spontaneous. Estimate the cell *emf* and free energy. The standard potential of (Fe^{2+}/Fe) = -0.41 V vs. NHE.

Ans39. Pt is inert and unlikely to participate in the redox reactions and will serve only for the electron transfer. Because the potential of (Fe^{+2}/Fe) is lower than that of (H^+/H_2 = 0 V), (H^+/H_2) will ensure the reduction half-reaction and (Fe^{+2}/Fe) the oxidation half-reaction.

Oxidation: $Fe \rightarrow Fe^{2+} + 2e^-$, $E^o = +0.41$ V

Reduction: $2H^+ + 2e^- \rightarrow H_2$, $E^o = 0$ V

Overall cell: $Fe + 2H^+ \rightarrow Fe^{2+} + H_2$

Note that the potential of Fe-based reaction is reversed because it is written in the oxidation form.

The cell *emf* is the sum of potentials of the two half-reactions written in their current states: *emf* = 0 + 0.41 = 0.41 V

The free energy $\Delta G = -nFemf = -2 \times 96.5 \times 0.41 = -79.13$ kJ mol^{-1}

The negative sign of the free energy confirms that the overall reaction is spontaneous in this direction.

Q40. Consider the following metal rods: Fe, Cd, Cu, and Au. Each rod is immersed in a cell containing acidic solution. Which of the metal rods will produce H_2 bubbles? Explain why using redox reactions. The standard potentials are: E^o (Fe^{2+}/Fe) = -0.44 V vs. NHE, E^o (Cd^{2+}/Cd) = -0.4 V vs. NHE, E^o (Au^{3+}/Au) = 1.5 V vs. NHE, and E^o (Cu^{2+}/Cu) = 0.337 V vs. NHE.

Ans40. Hydrogen bubbles could be produced from the reduction of protons in the acidic medium according to the reaction:

$2H^+ + 2e^- \rightarrow H_2$, $E^o = 0$ V

The standard potential of this electrode is zero. Therefore, as long as the oxidation potential of the other metal is superior to zero, etching of the metal in the acidic solution should produce hydrogen gas bubbles.

For iron, the reduction potential of the couple is (Fe^{2+}/Fe = -0.44 V, $Fe^{2+} + 2e^- \rightarrow Fe$), meaning that the oxidation form (Fe $\rightarrow Fe^{2+} + 2e^-$) has a potential of +0.44 V, which is superior to that of the hydrogen couple ($2H^+ + 2e^- \rightarrow H_2$, $E^o = 0$ V). Therefore, immersion of an Fe rod in acidic solution should induce etching of the metal by the protons and simultaneously the protons will reduce to form hydrogen gas.

For Cd rod (Cd^{2+}/Cd = -0.4 V), the same scenario as Fe should occur because of the similarity in the potentials values.

For the two remaining redox couples (Au^{3+}/Au = 1.5 V and Cu^{2+}/Cu = 0.337 V), hydrogen bubbles will not produce at these materials when immersed in acidic solutions because of their negative oxidation potentials compared to that of the hydrogen electrode.

The reduction potential of ($Au^{3+} + 3e^- \rightarrow Au$) is 1.5 V. Thus, the oxidation reaction ($Au \rightarrow Au^{3+} + 3e^-$) will have a potential of -1.5 V, which is lower than that of (H^+/H_2, 0 V).

The same scenario applies to the Cu rod (Cu^{2+}/Cu = 0.337 V). In other words, immersion of a Cu rod in acidic solution will not etch Cu and no hydrogen bubbles will be produced.

The contact between Fe and HCl should be summarized as follows:

Oxidation: $Fe \rightarrow Fe^{2+} + 2e^-$

Reduction: $2H^+ + 2e^- \rightarrow H_2$

Overall reaction: $Fe + 2H^+ \rightarrow Fe^{2+} + H_2$

Cl^- is a spectator ion and could be added to both sides of the reaction to yield:

$Fe + 2HCl \rightarrow FeCl_2 + H_2$

The products issued from the overall reactions (Fe and Cd) should be metal salts and hydrogen gas.

Q41. Calculate the masses of produced Na and Cl_2 when 10 F charge is passed through an electrolytic cell filled with molten NaCl.

Ans41. The electrolysis reaction of molten NaCl subjected to 10 F electrical charge is:

Overall reaction: $NaCl \rightarrow Na^+ + \frac{1}{2}Cl_2$

This overall reaction is composed of two half-reactions:

Oxidation: $Cl^- \rightarrow \frac{1}{2}Cl_{2(g)} + e^-$

Reduction: $Na^+ + e^- \rightarrow Na_{(molten)}$

The overall reaction indicates that passage of 1 mole of electrons (or 1 F = 96500 C) produces 1 mole of Na and $\frac{1}{2}$ moles of Cl_2. As 1 mole of sodium weights 22.98g and 1 mole of Cl_2 weights 35.45 × 2 = 70.9 g, the passage of 1 mole of electrons will produce 22.98 g of Na and 35.45 of Cl_2. If 10 F (or 10 moles of electrons) are passed through the cell, the masses should be multiplied by a factor of 10. This produces 229.8 g of Na and 354.5 g of Cl_2.

This could also be directly calculated using the Faraday's law of electrolysis:

$$q = it = \frac{mFz}{M}, \text{ or } m = \frac{qM}{Fz}$$

where m is the mass of the substance in grams, q is the total electrical charge passed through the substance, i is the electrical current in ampere (A), F is the Faraday constant ($F = 96485$ C mol^{-1}), M is the molar mass of the substance in (g mol^{-1}), and z represents the valence of the substance or number of the electrons transferred per ion.

Q42. An $AlCl_3$ molten salt is subjected to electrolysis. Calculate the number of Farads and the time required to deposit 50 Kg of Al metal.

Ans42. Before determining any quantities, it is advised to write down the balanced reactions to allow proper conversions of moles. The electrolysis of $AlCl_3$ molten salt will produce Al metal at one electrode and Cl_2 gas at the other.

Oxidation: $Cl^- \rightarrow \frac{1}{2} Cl_{2(g)} + e^-$

Reduction: $Al^{3+} + 3e^- \rightarrow Al_{(molten)}$

The equations show unbalances in the number of electrons, and both reactions should exchange 1 mole of electrons to facilitate the calculations.

Oxidation: $Cl^- \rightarrow \frac{1}{2} Cl_{2(g)} + e^-$

Reduction: $\frac{1}{3} Al^{3+} + e^- \rightarrow \frac{1}{3} Al_{(molten)}$

Therefore, the passage of 1 mole of electrons (or 1F) should produce $\frac{1}{2}$ moles of Cl_2 gas and $\frac{1}{3}$ moles of Al. On the other hand, 1 mole of Al weights 26.98 g. Hence, the deposition of 50000g will require $(\frac{50000}{26.98}) \times 3 = 5559.67$ moles of electrons.

This could also be directly calculated using the Faraday's law of electrolysis:

$$q = it = \frac{mFz}{M}, \text{ or } m = \frac{qM}{Fz}$$

where m is the mass of the substance in grams, q is the total electric charge passed through the substance, i is the electrical current in ampere (A), F is the Faraday constant ($F = 96485$ C mol^{-1}), M is the molar mass of the substance in (g mol^{-1}), and z represents the valence of the substance or number of the electrons transferred per ion.

Q43. Consider an electrochemical cell filled with Ag^+ solution containing a metal rod with an initial mass of 10 g. If 10 A current is applied for 2 min, by how much the electrode mass will increase?

Ans43. The application of an electric current to the solution of Ag^+ will reduce Ag^+ into Ag metal, which will deposit on the metal rod.

The involved reaction is: $Ag^+ + 1e^- \rightarrow Ag$

Using the Faraday's law of electrolysis, it is possible to estimate the mass.

$q = it = \frac{mFz}{M}$, or $m = \frac{qM}{Fz}$

where m is the mass of the substance in grams, q is the total electric charge passed through the substance, i is the electrical current in ampere (A), F is the Faraday constant (F = 96485 C mol^{-1}), M is the molar mass of the substance in (g mol^{-1}), and z represents the valence of the substance or number of the electrons transferred per ion.

This gives: $q = it = 10 \times 120 = 1200$ C

$m = \frac{qM}{Fz} = \frac{1200 \times 107}{96485 \times 1} = 1.33$ gr

Q44. Calculate the standard free energy of the Daniell galvanic cell (Zn/Cu) generating a voltage of 1.1 V.

Ans44. In the Daniell cell, 2 electrons are transferred during the redox reactions (n = 2).

$\Delta G = -nFE = -2 \times 96.5 \times 1.1 = -212.3$ kJ mol^{-1}

The negative sign of ΔG means that the reaction occur spontaneously to generate energy.

Q45. Consider a Zn/Ni cell with the standard free energy of -102 kJ mol^{-1} and the transfer of 2 electrons. Estimate the cell voltage. Are there other methods to calculate this cell voltage?

Ans45. The cell involves the transfer of 2 electrons from Zn to Ni.

$\Delta G = -nFE$, or $E = -\frac{\Delta G}{nF} = -\frac{(-102)}{96.5 \times 2} = 0.53$ V

This cell voltage can also be calculated using the standard potentials found in thermodynamic tables.

Oxidation: $Zn \rightarrow Zn^{2+} + 2e^-$, $E° = +0.763$ V

Reduction: $Ni^{2+} + 2e^- \rightarrow Ni$, $E° = -0.25$ V

Overall reaction: $Zn + Ni^{2+} \rightarrow Zn^{2+} + Ni$

The cell potential is the sum of potentials of both half-reactions written in their current states:

$E_{cell} = 0.763 + (-0.25) = 0.53$ V

Q46. Consider the following overall reaction:

$Zn^{2+}_{(aq)} + Pb_{(s)} \rightarrow Zn_{(s)} + Pb^{2+}_{(aq)}$

Calculate the overall cell voltage (or *emf*) at the standard conditions. The standard potentials are: E^o (Zn^{2+}/Zn) = -0.76 V vs. NHE and E^o (Pb^{2+}/Pb) = -0.126 V vs. NHE.

Ans46. This overall reaction is composed of two half-reactions:

Oxidation: $Pb_{(s)} \rightarrow Pb^{2+}_{(aq)} + 2e^-$, E^o = +0.126 V

Reduction: $Zn^{2+}_{(aq)} + 2e^- \rightarrow Zn_{(s)}$, E^o = -0.76 V

Overall reaction: $Zn^{2+}_{(aq)} + Pb_{(s)} \rightarrow Zn_{(s)} + Pb^{2+}_{(aq)}$

Note that the Pb-based potential is reversed since the reaction is written in the oxidation form. The cell voltage (or *emf*) is the sum of potentials of both half-reactions written in their current states:

E^o_{cell} = *emf* = 0.126 + (-0.76) = -0.634 V

Keep in mind that the oxidation and reduction half-reactions could also be determined using the oxidation number method if the overall reaction is provided. Since the oxidation state of Zn decreased from +2 to 0, it will ensure the reduction half-reaction. The oxidation number of Pb increased from 0 to +2, hence it will ensure the oxidation half-reaction.

Q47. A cell is constructed by two half-cells. The first consists of a Zn wire dipped in $Zn(NO_3)_2$ solution. The other half-cell is composed of an inert electrode immersed in Fe^{2+}/Fe^{3+} solution. Calculate the cell potential at 25 °C and the following concentrations: (Zn^{2+}) = 0.22 M, (Fe^{2+}) = 0.42 M, and (Fe^{3+}) = 0.69 M. The standard potentials are: E^o (Zn^{2+}/Zn) = -0.763 V vs. NHE and E^o (Fe^{3+}/Fe^{2+}) = +0.771 V vs. NHE.

Ans47. To correctly calculate the potential, balanced electrochemical reactions should first be written. Since Fe^{3+}/Fe^{2+} has a higher potential, it will ensure the reduction half-reaction and Zn^{2+}/Zn with lower potential will ensure the oxidation half-reaction.

Oxidation: $Zn \rightarrow Zn^{2+} + 2e^-$, E^o = +0.763 V

Reduction: ($Fe^{3+} + 1e^- \rightarrow Fe^{2+}$) × 2 , E^o = +0.771 V

Overall reaction: $2Fe^{3+} + Zn \rightarrow 2Fe^{2+} + Zn^{2+}$

Note that the potential of the Zn-based reaction is reversed since the reaction is written in the oxidation form. Also, to eliminate the electrons from the overall reaction, the reduction based-reaction is multiplied by a factor of 2. The overall cell potential at the standard conditions is simply the sum of potentials of both half-reactions written in their current states.

E^o_{cell} = *emf*^{o} = 0.771 + 0.763 = 1.534 V

At condition different from the standard (different concentrations in this case), the Nernst equation is applicable.

$$E = E^o - \frac{RT}{nF} \ln Q = E^o - \frac{RT}{nF} \ln \frac{(Zn^{2+})(Fe^{2+})^2}{(Fe^{3+})^2} = 1.569 \text{ V}$$

Q48. Calculate the quantity of Farads and Coulombs of electricity needed to reduce 0.11 g of Cu^{2+} to metallic copper. The molar mass of Cu = 63.54 g mol^{-1}.

Ans48. The reduction of Cu^{2+} into metallic copper (Cu) could be expressed by the reaction:

Reduction: $Cu^{2+} + 2e^- \rightarrow Cu$

The reaction indicates that 1 mole of Cu^{2+} gives 1 mole of Cu.

Using the Faraday's law of electrolysis, it is possible to estimate the electric charge.

$$q = it = \frac{mFz}{M}$$

where m is the mass of the substance in grams, q is the total electric charge passed through the substance, i is the electrical current in ampere (A), F is the Faraday constant (F = 96485 C mol^{-1}), M is the molar mass of the substance in (g mol^{-1}), and z represents the valence of the substance or number of the electrons transferred per ion.

This gives: $q = \frac{0.11 \times 96485 \times 2}{63.54} = 334.12$ C

The quantity of Farads required is: $\frac{334.12}{96485} = 0.0034$ F

Q49. Calculate the quantity of electricity needed to reduce 0.8 moles of iron(III) into iron(II).

Ans49. The reduction of Fe^{3+} into Fe^{2+} can be expressed by the reaction:

Reduction: $Fe^{3+} + 1e^- \rightarrow Fe^{2+}$

Using the Faraday's law of electrolysis, it is possible to estimate the quantity of electricity.

$$q = it = \frac{mFz}{M}$$

where m is the mass of the substance in grams, q is the total electric charge passed through the substance, i is the electrical current in ampere (A), F is the Faraday constant (F = 96485 C mol^{-1}), M is the molar mass of the substance in (g mol^{-1}), and z represents the valence of the substance or number of the electrons transferred per ion.

Note that $\frac{m}{M}$ represents the number of moles (n).

This gives: $q = nFz = 0.8 \times 96485 \times 1 = 77188$ C

The quantity of Farads required is: $\frac{77188}{96485} = 0.8$ F

Q50. Calculate the number of moles of Al formed by passing a current of 1.50 A through a molten salt of AlCl$_3$ for 9 hours.

Ans50. The electrolysis of the molten salt AlCl$_3$ (Al^{3+}, 3Cl$^-$) forms Al metal and chlorine gas according to the redox reactions:

Oxidation: (Cl$^-$ → $\frac{1}{2}$Cl$_{2(g)}$ + 1e$^-$) × 3

Reduction: Al^{3+} + 3e$^-$ → Al

Overall reaction: 3Cl$^-$ + Al^{3+} → $\frac{3}{2}$Cl$_2$ + Al

Note that the oxidation reaction is multiplied by a factor of 3 to eliminate the number of electrons in the overall reaction.

Using the Faraday's law of electrolysis, it is possible to estimate the number of moles.

$$q = it = \frac{mFz}{M}$$

where m is the mass of the substance in grams, q is the total electric charge passed through the substance, i is the electrical current in ampere (A), F is the Faraday constant (F = 96485 C mol^{-1}), M is the molar mass of the substance in (g mol^{-1}), and z represents the valence of the substance or number of the electrons transferred per ion.

Note that $\frac{m}{M}$ represents the number of moles (n).

This gives: $it = nFz$, or $n = \frac{it}{Fz} = \frac{1.5 \times 9 \times 60 \times 60}{96485 \times 3} = 0.1678$ moles

Q51. The quantity of electric charge passed through a circuit could be estimated by measuring the mass of the solid Ag deposited by electrolysis of dissolved Ag$^+$. Calculate the quantity of charge if the mass of the electrode increased by 0.298 g. The molar mass of Ag is 107.86 g mol^{-1}.

Ans51. The deposition of Ag could be expressed by the reduction reaction:

Reduction: Ag$^+$ + 1e$^-$ → Ag

Using the Faraday's law of electrolysis, it is possible to estimate the quantity of charge.

$$q = it = \frac{mFz}{M}$$

where m is the mass of the substance in grams, q is the total electric charge passed through the substance, i is the electrical current in ampere (A), F is the Faraday constant (F = 96485 C mol^{-1}), M is the molar mass of the substance in (g mol^{-1}), and z represents the valence of the substance or number of the electrons transferred per ion.

This gives: $q = \frac{0.298 \times 96485 \times 1}{107.86} = 266.88$ C

Q52. Calculate the time required to liberate 0.03 moles of hydrogen gas during electrolysis of HCl solution at the current of 0.063 A.

Ans52. The hydrogen evolution during electrolysis of HCl can be expressed by the reactions:

Oxidation: $Cl^- \rightarrow \frac{1}{2}Cl_{2(g)} + 1e^-$

Reduction: $H^+ + 1e^- \rightarrow \frac{1}{2}H_2$

Overall reaction: $Cl^- + H^+ \rightarrow \frac{1}{2}Cl_2 + \frac{1}{2}H_2$

Using the Faraday's law of electrolysis, it is possible to estimate the electrolysis time.

$q = it = \frac{mFz}{M}$, or $t = \frac{mFz}{iM}$

where m is the mass of the substance in grams, q is the total electric charge passed through the substance, i is the electrical current in ampere (A), F is the Faraday constant ($F = 96485$ C mol^{-1}), M is the molar mass of the substance in (g mol^{-1}), and z represents the valence of the substance or number of the electrons transferred per ion.

Note that $\frac{m}{M}$ represents the number of moles (n).

This gives: $t = \frac{nFz}{i}$

The overall reaction indicates that 1 mole of HCl liberates $\frac{1}{2}$ moles of H_2.

$t = \frac{2nFz}{i} = \frac{2 \times 0.03 \times 96485 \times 1}{0.063} = 91905$ seconds $= 25.52$ hours

Q53. The electrolysis of molten NaCl yields Na and Cl_2 while electrolysis of aqueous NaCl forms H_2 and Cl_2. Explain the difference using redox reactions.

Ans53. Molten salts like NaCl at 801 °C subjected to an electric current in a polarized cell will split into chlorine gas (Cl_2) at one electrode and sodium metal at the other electrode.

Oxidation: $Cl^- \rightarrow \frac{1}{2}Cl_{2(g)} + e^-$

Reduction: $Na^+ + e^- \rightarrow Na_{(molten)}$

Overall reaction: $Cl^- + Na^+ \rightarrow \frac{1}{2}Cl_2 + Na$

On the other hand, dissolved NaCl salt in water subjected to electrolysis should induce hydrogen gas (H_2) at the cathode and chlorine gas (Cl_2) at the anode.

Oxidation: $Cl^- \rightarrow \frac{1}{2}Cl_{2(g)} + e^-$

Reduction: $2H^+_{(aq)} + 2e^- \rightarrow H_{2(g)}$ or $2H_2O_{(aq)} + 2e^- \rightarrow H_{2(g)} + 2OH^-_{(aq)}$

The presence of several competing ions in solution should reduce or oxidase first those with higher affinities to electrons (lower energies or potentials). For example, Na^+ might be reduced to Na in an aqueous solution containing NaCl salt but this requires elevated energy (or potential), and because others species (e.g., H^+, H_2O) are present and require lower energies, they will rather be the species subjected to reduction. By comparison, in the molten state, only Na^+ and Cl^- are present in the medium. As a result, they will be the only species subjected to reduction and oxidation despite their higher energies (or potentials).

Q54. A quantity of electricity is passed through silver nitrate aqueous solution to deposit 3.50 g of silver on the cathode. Calculate the quantity of lead that will deposit if the same quantity of electricity is applied to $PbCl_2$ solution. The molar mass of Ag = 107.96 g mol^{-1} and that Pb = 207.2 g mol^{-1}.

Ans54. Using the Faraday's law of electrolysis, it is possible to estimate the deposited mass.

$$q = it = \frac{mFz}{M}, \text{ or } m = \frac{qM}{Fz}$$

where m is the mass of the substance in grams, q is the total electric charge passed through the substance, i is the electrical current in ampere (A), F is the Faraday constant (F = 96485 C mol^{-1}), M is the molar mass of the substance in (g mol^{-1}), and z represents the valence of the substance or number of the electrons transferred per ion.

The reduction of Ag can be expressed by the reaction:

Reduction: $Ag^+ + 1e^- \rightarrow Ag$

The quantity of electricity: $q = \frac{3.5 \times 96485 \times 1}{107.86} = 3131.37$ C

This same quantity of electricity is then passed through a Pb^{2+} solution to deposit Pb according to the reaction:

Reduction: $Pb^{2+} + 2e^- \rightarrow Pb$

$$m = \frac{qM}{Fz} = \frac{3131.37 \times 207.2}{96485 \times 2} = 3.36 \text{ g}$$

In sum, *3.36 g* of Pb will deposit of the cathode.

Q55. Molten $ZnCl_2$ is subjected to electrolysis at current of 1.7 A to deposit 25.0 g of Zn on the cathode. Estimate the mass of the liberated Cl_2 at the anode. The molar mass of Zn = 65.38 g mol^{-1} and that of Cl = 35.45 g mol^{-1}.

Ans55. The electrolysis of $ZnCl_2$ will generate Cl_2 at the anode and deposit Zn at the cathode according to the reactions.

Anode: $2Cl^- \rightarrow Cl_2 + 2e^-$

Cathode: $Zn^{2+} + 2e^- \rightarrow Zn$

Overall reaction: $2Cl^- + Zn^{2+} \rightarrow Cl_2 + Zn$

Using the Faraday's law of electrolysis, it is possible to estimate the electric charge and deposited mass.

$$q = it = \frac{mFz}{M}, \text{ or } m = \frac{qM}{Fz}$$

where m is the mass of the substance in grams, q is the total electric charge passed through the substance, i is the electrical current in ampere (A), F is the Faraday constant (F = 96485 C mol^{-1}), M is the molar mass of the substance in (g mol^{-1}), and z represents the valence of the substance or number of the electrons transferred per ion.

The quantity of electricity passed through the electrolysis cell is:

$$q = \frac{25 \times 96485 \times 1}{65.38} = 73799.32 \, C$$

This same quantity of electricity is passed through the anode to generate Cl$_2$.

$$m = \frac{qM}{Fz} = \frac{73799.32 \times 35.47}{96485 \times 2} = 27.12 \, g$$

Therefore, the electrolysis will generate 27.12 g of Cl$_2$.

Q56. The aim for building large dams on main Rivers is to provide cheap hydroelectric power for the production of metal, such as aluminum (Al). If the power plant at each dam induces a current of 10^8 A of electricity at voltage high enough to decompose molten aluminum salt, calculate the daily production of metallic Al if all electricity from one dam is used. How many dams would be needed for a daily production of 2500 metric tons (1 metric ton = 1000 kg) of Al?

Ans56. The production of Al could be expressed by the reduction reaction:

Reduction: $Al^{3+} + 3e^- \rightarrow Al$

Using the Faraday's law of electrolysis, it is possible to estimate the deposited mass.

$$q = it = \frac{mFz}{M}, \text{ or } m = \frac{itM}{Fz}$$

where m is the mass of the substance in grams, q is the total electric charge passed through the substance, i is the electrical current in ampere (A), F is the Faraday constant (F = 96485 C mol^{-1}), M is the molar mass of the substance in (g mol^{-1}), and z represents the valence of the substance or number of the electrons transferred per ion

For 24 hours (daily) production, $m = \frac{10^8 \times 24 \times 60 \times 60 \times 26.98}{96485 \times 3} = 8.05 \times 10^8 \, g = 8.05 \times 10^5 \, kg = 8.05 \times 10^2$ metric tons

The production of 2500 metric tons requires an average of 3 dams.

Q57. Potassium dichromate ($K_2Cr_2O_7$) is often employed in redox titrations. A solution of $K_2Cr_2O_7$ is prepared by adding 2.5 g of the salt into 100 ml of water. Calculate the molarity of the solution. If Cr^{3+} reduces during titration, calculate the normality of the solution. The molar mass of $K_2Cr_2O_7 = 294.185$ g mol^{-1}.

Ans57. The molarity is defined by: $\dfrac{\text{mass}}{\text{molar mass} \times \text{volume}} = \dfrac{2.5 \text{ g}}{294.185 \text{ g } mol^{-1} \times 0.1 \text{ L}} = 0.085$ mol L^{-1}

The normality involves the number of equivalents of the electrons transferred during the reaction. This could be estimated by calculating the change in oxidation number. The reduction half-reaction of $Cr_2O_7^{2-}$ could be summarized as follows:

$Cr_2O_7^{2-} + 14H^+ + 6e^- \rightarrow 2Cr^{3+} + 7H_2O$

The oxidation number changed from +6 in $Cr_2O_7^{2-}$ to +3 in Cr^{3+}. Therefore, 3 electrons are involved for Cr and 6 electrons for 2Cr.

Normality = Molarity × Equivalent of electrons = $0.085 \times 6 = 0.51$ N

Q58. Calculate the equivalent weight of an oxidizing agent able to oxidize Fe^{2+} into Fe^{3+} if 0.55 g of the oxidant needed 25 ml of 0.64 M Fe^{2+} solution. By comparing with data from the periodic table, identify the oxidant.

Ans58. The oxidation reaction can be expressed by:

Oxidation: $Fe^{2+} \rightarrow Fe^{3+} + 1e^-$

At the equivalent point: $C_{oxidant} \times V_{oxidant} = n_{oxidant} = \left(\dfrac{m}{M}\right)_{oxidant} = C_{reductant} \times V_{reductant}$

where n is the number of moles, m is the mass, and M is the molar mass of the oxidant.

$M_{oxidant} = \dfrac{m_{oxidant}}{C_{reductant} \times V_{reductant}} = \dfrac{0.55}{0.64 \times 0.025} = 34.37$ g mol^{-1}

Comparison of this value with molar masses of the periodic table indicates that the element could be chlorine Cl, and the slight difference could be due to errors in the measurements.

Q59. Calculate the number of moles of H_2SO_4 neutralizing 30 ml of 0.15 N HI according to the reaction:

$H_2SO_4 + HI \rightarrow H_2S + I_2 + 4H_2O$

Make sure that the reaction is mass and charge balanced.

Ans59. The reaction is not balanced, and by multiplying HI by 8 and I_2 by 4, it becomes balanced.

$H_2SO_4 + 8HI \rightarrow H_2S + 4I_2 + 4H_2O$

At neutralization: $N_{H2SO4} \times V_{H2SO4} = n_{H2SO4} = N_{HI} \times V_{HI}$

The stoichiometry of the reaction indicates that 1 mole H_2SO_4 reacts with 8 moles HI.

Therefore, n_{H2SO4} = 0.15 × 0.03 × $(\frac{1}{8})$ = 0.00056 moles

Q60. Determine the reduced/oxidized species and the reductant/oxidant in each of the following unbalanced reactions.

$6H^+ + MnO_4^- + 5SO_3^{2-} \rightarrow 5SO_4^{2-} + 2Mn_2 + 3H_2O$

$3Cl_2 + 6OH^- \rightarrow ClO^{3-} + 5Cl^- + 3H_2O$

Ans60. Each overall reaction is composed of two half-reactions. The oxidation number method could be used to determine the reduced/oxidized species and reductant/oxidant.

For the reaction: $6H^+ + MnO_4^- + 5SO_3^{2-} \rightarrow 5SO_4^{2-} + 2Mn_2 + 3H_2O$

The oxidation number of Mn decreases from +7 in MnO_4^- to 0 in Mn_2. Thus, MnO_4^- is the species reduced to Mn_2 and SO_3^{2-} is oxidized to SO_4^{2-}. MnO_4^- is the oxidant and SO_3^{2-} is the reductant.

For the reaction: $3Cl_2 + 6OH^- \rightarrow ClO^{3-} + 5Cl^- + 3H_2O$

The oxidation number of Cl increased from 0 in Cl_2 to +6 in ClO^{3-}, and decreased from 0 in Cl_2 to -1 in Cl^-. Hence, Cl_2 is oxidized to ClO^{3-}, and Cl_2 is reduced to Cl^-. In this case, Cl_2 plays the role of both the oxidant and reductant to give different reductant and oxidant.

Q61. A reducing agent is titrated by 20 mL of 0.5 g I_2 as the oxidant to transform I_2 into I^-. Estimate the molarity and normality of the reducing agent solution.

Ans61. The molarity is defined by:

$$\frac{number\ of\ moles}{volume} = \frac{mass}{molar\ mass \times volume} = \frac{0.5}{(126.9 \times 2) \times 20 \times 10^{-3}} = 0.0985 \text{ mol L}^{-1}$$

The reduction of I_2 consumes 2 electrons according to the reaction:

Reduction: $I_2 + 2e^- \rightarrow 2I^-$

The oxidation number of I changes from 0 in I_2 to -1 in I^-. Therefore, the number of the transferred electrons is $2e^-$. In other words, the reduction of I_2 involves 2 moles of electrons.

The normality is calculated as: molarity × number of mole of electrons = 0.0985 × 2 = 0.197 equivalent liter^{-1}

Q62. Explain the Daniell Cell through redox reactions. The standard potentials are: E^o (Zn^{2+}/Zn) = -0.76 V vs. NHE and E^o (Cu^{2+}/Cu) = 0.34 V vs. NHE.

Ans62. The Daniell cell is composed of two pots: one composed of a Zn rod immersed in $ZnSO_4$ solution and the other contains a Cu rod immersed in $CuSO_4$ solution. To ensure an electron flow through the external circuit, the rods are connected by a conducting wire. The ion transport in the internal circuit is ensured by a salt bridge. The electrons flow from Zn to Cu. Since the potential of (Zn^{2+}/Zn) is lower than that of (Cu^{2+}/Cu), oxidation will occur at the Zn pole and reduction at the Cu pole.

Oxidation: $Zn_{(s)} \rightarrow Zn^{2+}_{(aq)} + 2e^-$

Reduction: $Cu^{2+}_{(aq)} + 2e^- \rightarrow Cu_{(s)}$

Overall reaction: $Zn_{(s)} + Cu^{2+}_{(aq)} \rightarrow Zn^{2+}_{(aq)} + Cu_{(s)}$

Q63. Calculate the voltage of a cell composed of a Cu rod immersed in 1M $CuSO_4$ connected to a hydrogen electrode immersed in 1M HCl solution at 25°C. The standard potential of (Cu^{2+}/Cu) = 0.34 V vs. NHE.

Ans63. Note that the provided conditions are the standard: temperature of 25°C, concentration of 1 mol L^{-1}, and pressure of 1 Atm. Therefore, there is no need for using the Nernst equation but the standard potentials suffice to estimate the cell voltage (or $emf°$). The two half-reactions and overall reaction can be summarized as follows:

Oxidation: $H_{2(g)} \rightarrow 2H^+ + 2e^-$, $E° = 0.00$ V

Reduction: $Cu^{2+} + 2e^- \rightarrow Cu_{(s)}$, $E° = 0.34$ V

Overall reaction: $Cu^{2+} + H_{2(g)} \rightarrow Cu_{(s)} + 2H^+$

The standard potential of the hydrogen electrode is 0 V, and the potential of the overall reaction is the sum of potentials of the two half-reactions written in their current states: $emf° = 0 + 0.34 = 0.34$ V

Q64. Consider a cell with the following overall reaction:

$Au^+ + Cu \rightarrow Cu^{2+} + Au$

Write down the two half-reactions and balanced overall reaction. Calculate the cell potential at the standard conditions. The standard potentials are: $E°$ (Cu^{2+}/Cu) = +0.34 V vs. NHE and $E°$ (Au^+/Au) = +1.68 V vs. NHE.

Ans64. The overall reaction indicates that the oxidation number of Au decreases from +1 in Au^+ to 0 in Au. Hence, the Au terminal will ensure the reduction half-reaction. The oxidation state of Cu increases from 0 in Cu to +2 in Cu^{2+}, indicating that the oxidation half-reaction will occur at the Cu terminal.

Oxidation: $Cu \rightarrow Cu^{2+} + 2e^-$, $E° = -0.34$ V

Reduction: ($Au^+ + e^- \rightarrow Au$) × 2, $E° = +1.68$ V

Overall reaction: $2Au^+ + Cu \rightarrow Cu^{2+} + 2Au$

Note that potential of the Cu-based half-reaction is reversed because it is written in the oxidation form. Also, the reduction reaction is multiplied by a factor of 2 to eliminate the number of the electrons in the overall reaction.

The cell potential is the sum of potentials of both half-reactions written in their current states:

$emf^o = (-0.34) + 1.68 = +1.34$ V

Q65. Consider a cell with the following unbalanced overall reaction:

$Sn^{4+} + Ce^{3+} \rightarrow Sn^{2+} + Ce^{4+}$

i) Write down the two half-reactions and balanced overall reaction. ii) Calculate the cell potential (or *emf*) at the standard conditions. iii) Is the cell spontaneous? iv) Is it possible to estimate the cell voltage at higher temperatures and concentrations? The standard potentials are: E^o $(Ce^{4+}/Ce^{3+}) = +1.61$ V vs. NHE and $E^o (Sn^{4+}/Sn^{2+}) = +0.15$ V vs. NHE.

Ans65. i) The overall reaction indicates that the oxidation number of Sn decreases from +4 in Sn^{4+} to +2 in Sn^{2+}. Thus, the Sn terminal will ensure the reduction half-reaction. The oxidation state of Ce increases from +3 in Ce^{3+} to +4 in Ce^{4+}, indicating that the oxidation half-reaction will occur at the Ce terminal.

Oxidation: $(Ce^{3+} \rightarrow Ce^{4+} + e^-) \times 2$, $E^o = -1.61$ V

Reduction: $Sn^{4+} + 2e^- \rightarrow Sn^{2+}$, $E^o = +0.15$ V

Overall reaction $Sn^{4+} + 2Ce^{3+} \rightarrow Sn^{2+} + 2Ce^{4+}$

ii) Note that the potential of the Ce-based half-reaction is reversed because it is written in the oxidation form. Also, the oxidation reaction is multiplied by a factor of 2 to eliminate the number of electrons in the overall reaction.

The cell potential (or *emf*) is the sum of potentials of both half-reactions written in their current states: $emf^o = (-1.61) + 0.15 = -1.46$ V

iii) Since the cell potential is negative, the Gibbs free energy is positive ($\Delta G = -nFE$), meaning that the reaction is not spontaneous.

iv) Yes, it is possible to estimate the cell potential at conditions different from the standard by using the Nernst equation: $E = emf = emf^o - \frac{RT}{nF} \ln Q$, where Q is the reaction quotient expressing the concentrations of the redox species.

Q66. Consider the overall reaction: $AgCl \rightarrow Ag^+ + Cl^-$

Identify the two redox half-reactions and estimate the solubility constant of AgCl at the standard conditions. The standard potentials are: E^o (Ag^+/Ag) = +0.8 V vs. NHE and E^o (AgCl/Ag) = +0.22 V vs. NHE.

Ans66. The overall reaction indicates that the oxidation number of Ag decreases from +1 in AgCl to 0 in Ag. Thus, it will ensure the reduction half-reaction. The oxidation number of Ag increases from 0 in Ag to +1 in Ag^+, suggesting that the oxidation half-reaction will occur at this terminal.

Oxidation: $Ag \rightarrow Ag^+ + 1e^-$, E^o = -0.8 V

Reduction: $AgCl + 1e^- \rightarrow Ag + Cl^-$, E^o = +0.22 V

Overall reaction: $AgCl \rightarrow Ag^+ + Cl^-$

The solubility equilibrium constant of the reaction could be estimated by the Nernst equation:

$emf = emf^o - \frac{RT}{nF} \ln Q$, where Q is the reaction quotient expressing the concentrations (or activities) of the redox species.

At equilibrium, $Q = K_{eq} = K_s = \frac{(Ag^+)(Cl^-)}{(AgCl)}$, and the cell voltage (or emf) is 0 because $\Delta G = 0$.

Therefore, $emf^o = \frac{RT}{nF} \ln K_s = \frac{0.0592}{n} \log K_s$

This gives: $\log K_s = \left(\frac{1}{0.0592}\right) \times emf^o = \left(\frac{1}{0.0592}\right) \times (-0.8 + 0.22)$

In sum, $K_s = 1.6 \times 10^{-10}$

Q67. Consider the overall reaction:

$Hg_2Cl_2 \rightarrow Hg_2^{2+} + 2Cl^-$

The solubility constant of Hg_2Cl_2 is $K_s = 1.3 \times 10^{-18}$ and the standard potential of (Hg_2^{2+}/Hg) is +0.79V vs. NHE. Calculate the potential of the redox couple (Hg_2Cl_2/Hg).

Ans67. From the overall reaction, the solubility constant could be written as:

$K_s = \frac{(Hg_2^{2+})(Cl^-)^2}{(Hg_2Cl_2)}$

This overall reaction could be decomposed into oxidation and reduction half-reactions. The provided redox couples will help determining these reactions.

Oxidation: $2Hg \rightarrow Hg_2^{2+} + 2e^-$, E^o = -0.79 V

Reduction: $Hg_2Cl_2 + 2e^- \rightarrow 2Hg + 2Cl^-$, E^o = ? V

Note that oxidation and reduction could also be identified by calculating the change in oxidation numbers of the elements: increase = oxidation and decrease = reduction.

The Nernst equation states that: $emf = emf^o - \frac{RT}{nF} \ln Q$, where Q is the reaction quotient expressing the concentrations (or activities) of the redox species.

At equilibrium, $emf = 0$ because $\Delta G = 0$, and $Q = K_s$.

Therefore, $emf^o = \frac{RT}{nF} \ln K_s = \frac{0.0592}{2} \log(1.3 \times 10^{-18}) = -0.529$ V

In turn, the obtained emf is the sum of potentials of both half-reactions written in their current states. $emf^o = -0.529 = -0.79 + E_{red}$

This leads to: $E^o (Hg_2Cl_2/Hg) = +0.27$ V vs. NHE

Q68. Ag^+ in presence of CN^- forms a complex $Ag(CN)_2^-$. i) Write down the overall reaction showing the formation of the complex. ii) Divide the overall reaction into two redox half-reactions (oxidation and reduction). iii) Is it possible to estimate the concentration of CN^- if the equilibrium complexation constant (K_c) is known? iv) Estimate the emf and emf^o at equilibrium.

Ans68. i) The interaction between Ag^+ with CN^- forms a complex according to the overall complexation reaction:

$Ag^+ + 2CN^- \rightarrow Ag(CN)_2^-$

The complexation constant of this reaction is: $K_c = \frac{(Ag(CN)_2^-)}{(Ag^+)(CN^-)^2}$

ii) This overall reaction could be decomposed into oxidation and reduction half-reactions:

Oxidation: $Ag + 2CN^- \rightarrow Ag(CN)_2^- + 1e^-$

Reduction: $Ag^+ + 1e^- \rightarrow Ag$

iii) If the reaction performs to completion, the stoichiometry of the reaction indicates that 1 mole Ag^+ reacts with 2 moles CN^- to form 1 mole $Ag(CN)_2^-$. The equilibrium constant becomes:

$K_c = \frac{(CN^-)}{(CN^-)(2 \times (CN^-))^2} = \frac{1}{4(CN^-)^2}$

Therefore, the knowledge of the equilibrium constant should allow determining the concentration of CN^-.

iv) Using the Nernst equation, it is possible to estimate emf and emf^o.

$emf = emf^o - \frac{RT}{nF} \ln Q$, where Q is the reaction quotient.

At equilibrium, $emf = 0$ because Gibbs free energy equals to zero, and $Q = K_c$.

Therefore, $emf^o = \frac{RT}{nF} \ln K_c$

By knowing K_c and T, emf^o could be calculated ($n = 1$).

Q69. i) Consider the two redox couples involved in corrosion of Fe and Al with E^o (Fe^{2+}/Fe) = -0.41 V vs. NHE and E^o (Al^{3+}/Al) = -1.66 V vs. NHE. Which of the metals is easy to corrode and why? ii) By which means could the corrosion be prevented?

Ans69. i) The redox potentials indicate that Al is easier to oxidize than Fe because it has a lower potential. Therefore, in theory, Al is easy to corrode than Fe. However, the difference in crystal structures between Fe and Al contradict these assumptions. In fact, Al is highly resistant to corrosion when compared to Fe. The reason for this has to do with the formation of Al oxides during the first stages of corrosion, which have similar crystal structure as that of Al in terms of cell unit and packaging distance. As a result, the first formed layers of oxides during corrosion of Al deposit and adhere well to the Al surface, forming a protective layer which prevents further corrosion. By comparison, oxides formed during the first corrosion stages of Fe have crystal structures very different from that of Fe. Therefore, the oxides do not adhere properly to the surface of Fe but instead deposit as flacks. This, in turn, allows more moisture and oxygen to infiltrate to the remaining Fe and cause further corrosion.

ii) Several methods could be used to prevent or decelerate corrosion of metals, including painting, electroplating, and the use of sacrificial electrodes.

Highly adhesive paints could prevent corrosion of metals because they block moisture and oxygen from reaching the metal to initiate the corrosion process. However, in the long run, this often has its limitations as paint could detach from the metal and corrosion could happen. Continuous maintenance of the material is required over time.

The other method to protect from corrosion deals with depositing thin layers of non-corrosive metals (e.g., gold, platinum) on top of the corrosive metals (e.g., iron). In electroplating, the metal is often immersed in a bath containing dissolved salts of non-corrosive metals (e.g., AuCl$_3$, H$_2$PtCl$_6$). A cathodic potential is then applied to deposit the metal cations on the corrosive metal surface. This process is often more efficient than painting but it still has limitations as the deposited films could deteriorate over time. The thicker is the deposited layer, the better is the stability over time.

The third way to protect corrosive metals is by means of sacrificial electrodes (e.g., magnesium). For example, by connecting an Mg electrode to a piece of Fe, corrosion will be more focused on the Mg because of its lower potential (Mg^{2+} + 2e^- → Mg, -2.37 V vs. NHE).

This will protect Fe from corrosion. Maintenance of the system by replacing deteriorated Mg electrodes after severe corrosion is necessary to keep corrosion away from the other metal.

Q70. Compare the corrosion tendency of Fe to that of Al.

Ans70. Al has a greater tendency to resist corrosion than iron. During corrosion of Al, a layer of Al-oxides is formed on the surface of Al, which prevents further corrosion of Al. By contrast, corrosion of Fe forms Fe oxides/hydroxides flacks which are very porous to stop the corrosion process. Therefore, corrosion will continue and progress to deeper layers of Fe.

Q71. Estimate the *emf* of the cell: $2Al_{(s)} | 2Al^{3+}_{(aq)} || 3Fe^{2+}_{(aq)} | 3Fe_{(s)}$. The standard potentials are: $E^o(Al^{+3}/Al) = -1.66$ V vs. NHE and $E^o(Fe^{2+}/Fe) = -0.44$ V vs. NHE.

Ans71. The best method to correctly calculate the potential (or *emf*) is first to write down the two half-reactions with their corresponding potentials then sum these potentials. The cell diagram indicates that the oxidation occurs at the Al terminal and reduction at the Fe terminal.

Oxidation: $2Al_{(s)} \rightarrow 2Al^{3+}_{(aq)} + 6e^-$, $E^o = +1.66$ V

Reduction: $3Fe^{2+}_{(aq)} + 6e^- \rightarrow 3Fe_{(s)}$, $E^o = -0.44$ V

Overall reaction: $2Al_{(s)} + 3Fe^{2+}_{(aq)} \rightarrow 2Al^{3+}_{(aq)} + 3Fe_{(s)}$

Note that the Al-based potential is reversed because the reaction is written in the oxidation form. The cell potential (or *emf*) is the sum of potentials of both half-reactions written in their current states: *emf* = 1.66 + (-0.44) = 1.22 V

Q72. Consider an electrochemical cell composed of two half cells: the first is made of a Cu rod immersed in CuSO$_4$ (4M) and the other by an Mg rod immersed in MgSO$_4$ (1M). i) Could this cell eventually produce an *emf*? If so, what would be its origin? ii) Estimate the cell *emf* in dry weather of 35 °C. iii) Could this cell eventually be used in vehicles starting engines? If not, propose a suitable design for this application. The standard potentials are: $E^o(Mg^{2+}/Mg) = -2.37$ V vs. NHE and $E^o(Cu^{2+}/Cu) = 0.337$ V vs. NHE.

Ans72. i) Yes, this cell will produce a difference in potential. The origin of this *emf* is the difference in standard potentials between the Cu and Mg metals, as well as the difference in ion concentrations (4M and 1M). The standard potentials indicate that oxidation occurs at the Mg half-cell because of its low potential and reduction at the Cu half-cell.

Oxidation: $Mg \rightarrow Mg^{2+} + 2e^-$, $E^o = +2.37$ V

Reduction: $Cu^{2+} + 2e^- \rightarrow Cu$, $E^o = 0.337$ V

Overall reaction: $Mg + Cu^{2+} \rightarrow Mg^{2+} + Cu$

Note that the standard potential of the Mg-based half-reaction is reversed in sign because it is written in the oxidation form.

iii) The *emf* of the overall cell could be estimated by means of the Nernst equation.

$$emf = emf^o - \frac{RT}{nF} Ln\, Q = emf^o - \frac{RT}{nF} Ln\, Q\, \frac{(Mg^{2+})}{(Cu^{2+})} = (2.37 + 0.337) - \frac{8.31 \times 308.1}{2 \times 96485} Ln\, Q\, \frac{1}{4} = 2.72\ V$$

iii) This cell delivers a substantial voltage that could be used for vehicle starting engines. However, the cell will not be suitable for this application because the electrolytes are liquid, which will make it difficult to handle during vehicle motion. Another alternative to adopt this cell for vehicle application is to design a cell using these components but by replacing the $CuSO_4$ and $MgSO_4$ solutions with paste or solid electrolytes.

Q73. What is the origin of *emf* in concentration cell? Express the cell *emf* as a function of concentrations of the species.

Ans73. In concentration cell, the electrodes are made of the same material but the electrolytes in both half-cells have different concentrations. Therefore, the *emf* originates from the difference in concentration. The Nernst equation for concentration cell can be written as:

$$emf = emf^o - \frac{RT}{nF} Ln\, Q,$$ where Q represents the reaction quotient linked to concentrations (or activities) of the redox species. Concentration cells often generate very small voltages because $emf^o = 0$, and only the concentration gradient counts (Q).

Q74. AgCl is a slightly soluble salt in aqueous solution. Could electrochemistry be used to determine the concentration of Ag^+?

Ans74. Yes, electrochemistry can be used to determine the concentrations of slightly soluble salts like AgCl using the Nernst equation.

Solubility reaction: $AgCl \rightarrow Ag^+ + Cl^-$

$$emf = emf^o - \frac{RT}{nF} Ln\, Q = emf^o - \frac{RT}{nF} Ln\, (Ag^+)(Cl^-)$$

Q75. i) What would happen to an Fe rod exposed to wet conditions? Explain the phenomena using redox reactions. ii) How could this phenomenon be prevented? iii) Would the same phenomenon occur for an Al rod exposed to the same conditions?

Ans75. i) If an Fe rod is left under wet conditions, it will corrode after a certain time. The presence of moisture and oxygen (oxidant) in the air will attack the Fe surface and oxidation occurs to form Fe^{2+}.

Oxidation: $Fe \rightarrow Fe^{2+} + 2e^-$, $E^o = 0.41\ V$

Reduction: $½O_2 + H_2O + 2e^- \rightarrow 2OH^-$, $E° = 0.4$ V

Fe^{2+} will combine with OH^- to form oxides/hydroxides precipitates or complexes, appearing as flacks on the Fe rod.

ii) Several methods could be used to prevent the corrosion process from happening or at least decelerate its kinetics, including painting/adhesives, electroplating, and use of sacrificial electrodes.

Putting strong adhesive/paint on the Fe surface will prevent moisture and oxygen from attacking the Fe surface and prevent corrosion.

Electroplating consists of depositing thin layers of other metals with good resistance to corrosion, such as Ni. In this process, Fe is immersed in the metal salt of the noncorrosive metal and potential is applied to drive the metal cations and deposit on the Fe rod.

Sacrificial electrodes are another way to prevent corrosion. By coupling Fe with Mg, it will be possible to prevent corrosion of Fe because Mg will corrode first due to its very low potential compared to that of Fe. This process is often used to protect boats and ships from corrosion in salted waters of seas and oceans, where corrosion processes are further accelerated due to the high content in salts (high conductivity electrolytes).

iii) Al under wet conditions will not corrode as Fe due to the difference in crystal structures between Fe and Al. During the initial stages of corrosion, the top layer of Al will undergo corrosion to form oxides. However, because the crystal structure of the formed oxide is close to that of Al, the corrosion products will homogeneously cover the Al surface and form a protective layer, which will limit further corrosion.

Q76. i) In your opinion, what will happen if Zn powder (0.5 M) is suspended in $CuCl_2$ solution (pH = 0)? ii) Will you see any gas evolution during this reaction? iii) Explain why the setup will not generate an *emf*. iv) Propose a better design for using these components to produce an *emf*. The standard potentials are: $E°$ (Zn^{2+}/Zn) = -0.76 V vs. NHE and $E°$ (Cu^{+2}/Cu) = 0.34 V vs. NHE.

Ans76. i) The difference in potential between Cu and Zn will create some sort of galvanic cell. Zn powder is composed of Zn metal ground into fine particles. The suspension of these particles in acidic media will oxidize Zn because of its lower potential in presence of the reducing couple Cu^{2+}/Cu with higher potential. The following reactions will occur in acidic solution.

Oxidation: $Zn \rightarrow Zn^{2+} + 2e^-$

Reduction: $Cu^{2+} + 2e^- \to Cu$

ii) Because the solution is acidic, hydrogen evolution may also occur according to the reaction.

$2H^+ + 2e^- \to H_2$

However, because the potential of (Cu^{2+}/Cu = 0.337 V) is higher than that of (H^+/H_2 = 0 V), the Cu will be more likely to reduce than protons.

iii) The overall reaction is unlikely to produce any *emf* because no separator is used, and most of the produced current will be short-circuited.

iv) To extract voltage out of this cell, both the anode and cathode reaction should first be separated by salt bridge or membrane to prevent ions mixing. The cell could then be connected to an external circuit (or electric wiring) to let the electrons flow from one pole to the other. Because Zn powder is difficult to connect, it should first be pressed into a pellet and metal wire could be connected to the powder pellet to collect the electrons. The Zn powder pellet could then be immersed in HCl solution at the same (pH = 0.) At the reduction pole, a Cu rod could be immersed in the $CuCl_2$ solution to allow Cu^{2+} to reduce into Cu deposit.

Q77. Consider the cell with the following diagram: Ag│AgCl/Cl⁻ (10^{-2}M)││Cu^{2+} (2M)│Cu

i) Identify the two redox half-reactions and overall reaction. ii) Calculate the cell *emf* at 20°C. iii) In which direction will the electrons flow spontaneously? The standard potentials are: E^o (AgCl/Ag) = 0.22 V vs. NHE and E^o (Cu^{2+}/Cu) = 0.34 V vs. NHE.

Ans77. i) The cell diagram indicates that Ag pole is the anode and Cu is the cathode.

Oxidation: $2Ag + 2Cl^- \to 2AgCl + 2e^-$, E^o = -0.22 V

Reduction: $Cu^{2+} + 2e^- \to Cu$, E^o = 0.34 V

Overall reaction: $2Ag + 2Cl^- + Cu^{2+} \to 2AgCl + Cu$

ii) The electrons flow from the Ag pole where they are generated to the Cu pole where they will be consumed to deposit Cu.

iii) Using the Nernst equation:

$$emf = emf^o - \frac{RT}{nF} Ln\, Q = emf^o - \frac{RT}{nF} Ln\, \frac{1}{(Cu^{2+})(Cl^-)^2} = (-0.22 + 0.34) - \frac{8.31 \times 293.15}{2 \times 96485} Ln\, \frac{1}{(2)(10^{-2})^2} =$$

0.013 V

iii) The free energy of the cell is: ΔG = -nF*emf*. Because emf>0, ΔG<0, meaning that the reaction is spontaneous in this direction.

Q78. Consider the cell: H_2│OH^- (10^{-4} M)││Br^- (10^{-3} M)│Br_2

i) Identify the anode and cathode, and write down the two half-reactions and overall reaction. ii) Calculate the cell *emf* and free energy at 1 atm and 45 °C. iii) Is the reaction spontaneous? iv) Estimate the *emf* and ΔG at equilibrium. The standard potentials are: E^o (Br_2/Br^-) = 1.06 V vs. NHE and E^o (H_2O/H_2) = -0.86 V vs. NHE.

Ans78. i) The cell notation indicates that the H_2 half-cell is the anode and Br_2 is the cathode.

Oxidation: $H_2 + 2OH^- \rightarrow 2H_2O + 2e^-$, E^o = 0.86 V

Reduction: $Br_2 + 2e^- \rightarrow 2Br^-$, E^o = 1.06 V

Overall reaction: $H_2 + Br_2 + 2OH^- \rightarrow 2H_2O + 2Br^-$

ii) The potential can be estimated using the Nernst equation:

$$emf = emf^o - \frac{RT}{nF} \ln Q = emf^o - \frac{RT}{nF} \ln \frac{(Br^-)^2}{(OH^-)^2} = (0.86 + 1.06) - \frac{8.31 \times 318.15}{2 \times 96485} \ln \frac{(10^{-3})^2}{(10^{-4})^2} = 1.86 \text{ V}$$

iii) $\Delta G = -nF\,emf = -2 \times 96485 \times 1.86 = -358.94$ kJ mol^{-1}

The free energy is negative, meaning that the reaction will proceed spontaneously to generate electricity.

iv) When the equilibrium is reached, $\Delta G = 0$ and $emf = 0$.

Q79. Consider a cell composed of Sn|Sn^{2+}(0.15M) and Fe|Fe^{3+} (0.05 M) half-cells. i) Identify the direction of a spontaneous flow of electrons, and write down the two half-reactions. ii) Estimate the cell emf^o. iii) Is this cell useful in terms of energy? The standard potentials are: E^o (Sn^{2+}/Sn) = -0.14 V vs. NHE and E^o (Fe^{3+}/Fe) = -0.04 V vs. NHE.

Ans79. i) The redox reactions involved in this cell are:

$Sn^{2+} + 2e^- \rightarrow Sn$, E^o = -0.14 V

$Fe^{3+} + 3e^- \rightarrow Fe$, E^o = -0.04 V

ii) At the standard conditions, (Sn^{2+}/Sn) has a lower potential than (Fe^{3+}/Fe), meaning that the Sn terminal will ensure the oxidation half-reaction and Fe terminal the reduction half-reaction. However, since the concentrations are different from the standard (1 M), the potential order could be reversed. The best way to determine the flow of electrons under these conditions is by calculating both potentials and comparing the values.

The Nernst equation can be used for calculating potentials at conditions different from the standard.

$$E_{Sn} = E^0_{Sn} - \frac{RT}{nF} \ln \frac{1}{(Sn^{2+})} = -0.14 - \frac{8.31 \times 298.15}{2 \times 96485} \ln \frac{1}{(0.15)} = -0.164 \text{ V}$$

$$E_{Fe} = E^0_{Fe} - \frac{RT}{nF} Ln \frac{1}{(Fe^{3+})} = -0.04 - \frac{8.31 \times 298.15}{3 \times 96485} Ln \frac{1}{(0.05)} = -0.065 \text{ V}$$

Since Sn pole still has the lowest potential, it will ensure the oxidation half-reaction at the anode and the Fe pole with relatively higher potential will ensure the reduction half-reaction at the cathode. The redox reactions in this cell are:

Oxidation: $(Sn \rightarrow Sn^{2+} + 2e^-) \times 3$

Reduction: $(Fe^{3+} + 3e^- \rightarrow Fe) \times 2$

Overall reaction: $3Sn + 2Fe^{3+} \rightarrow 3Sn^{2+} + 2Fe$

To cancel the electrons in the overall reaction, each half-reaction is multiplied by the number of electrons of the other half-reaction. Therefore, electrons will flow from the Sn pole where are generated by oxidation to the Fe pole where they will be consumed to reduce Fe^{3+} into Fe.

iii) $emf^0 = 0.14 - 0.04 = 0.1$ V

$$emf = emf^0 - \frac{RT}{nF} Ln \frac{(Sn^{2+})^3}{(Fe^{3+})^2} = 0.1 - \frac{8.31 \times 298.15}{2 \times 96485} Ln \frac{(0.15)^3}{(0.05)^2} = 0.096 \text{ V}$$

iv) This cell delivers a very small voltage of 0.096 V, which would not be useful for most applications, except maybe in micro or nanoelectronics.

Q80. i) Calculate the *emf* of the following cell: $Cd\,|\,Cd^{2+}\,(10^{-7}\,M)\,||\,Pb^{2+}\,(10^7\,M)\,|\,Pb$ at 5 °C. ii) Estimate the cell free energy ΔG. iii) Is the reaction spontaneous? iv) Calculate the potential of each half-cell. v) Are both potentials additives? The standard potentials are: $E^o\,(Pb^{2+}/Pb) = -0.13$ V vs. NHE and $E^o\,(Cd^{2+}/Cd) = -0.4$ V vs. NHE.

Ans80. i) The cell diagram suggests that Cd immersed in Cd^{2+} is the anode and Pb immersed in Pb^{2+} is the cathode. The simplest way to determine the potential and electron flow is to write down the two half-reactions with their standard potentials (with the correct sign), then estimate the overall cell voltage.

Oxidation: $Cd \rightarrow Cd^{2+} + 2e^-$, $E^o = 0.4$ V

Reduction: $Pb^{2+} + 2e^- \rightarrow Pb$, $E^o = -0.13$ V

Overall reaction: $Cd + Pb^{2+} \rightarrow Cd^{2+} + Pb$

Note that the standard potential of Cd-based reaction is reversed in sign because it is written in the oxidation form.

Using the Nernst equation:

$$emf = emf^o - \frac{RT}{nF} Ln \frac{(Cd^{2+})}{(Pb^{2+})} = (0.4 - 0.13) - \frac{8.31 \times 278.15}{2 \times 96485} Ln \frac{(10^{-7})}{(10^7)} = 0.656 \text{ V}$$

ii) The free energy of the reaction is: $\Delta G = -nFemf = -2 \times 96485 \times 0.656 = -126.58$ kJ mol^{-1}

iii) The free energy is negative, meaning that the reaction occurs spontaneously.

iv) Nernst equation can also be used to estimate the potential of each half-cell separately.

$$E_{Cd} = E^0_{Cd} - \frac{RT}{nF} Ln \frac{(Cd^{2+})}{(Cd)} = 0.4 - \frac{8.31 \times 278.15}{2 \times 96485} Ln \frac{(10^{7-})}{1} = 0.59 \text{ V}$$

$$E_{Pb} = E^0_{Pb} - \frac{RT}{nF} Ln \frac{(Pb)}{(Pb^{2+})} = -0.13 - \frac{8.31 \times 278.15}{2 \times 96485} Ln \frac{1}{(10^7)} = 0.06 \text{ V}$$

v) Yes, it can be seen that the potentials are additives in their current written states since the overall potential is the sum of potentials of the two half-reactions.

Table of Content

Discount offers	1
Introduction	2
Abstract	3
1. Electrolytic cells	3
1.1. Electrolysis	3
1.2. Electroplating	5
2. Galvanic/voltaic cells (batteries)	5
2.1. Daniell Cell	6
2.2. Hydrogen electrode based cell	7
2.3. Weston Cell	8
2.4. Solid, moist cells or dry cells	8
2.5. Reversible lead storage batteries	9
2.6. Fuel cells	9
3. Redox titrations	10
4. Solubility, precipitation, and complexation reactions	10
5. Corrosion	11
6. Combustion	11
Summary	12
References	13
Practical Questions/Problems with Solutions	14
Table of content	56
About the author	58

www.ingramcontent.com/pod-product-compliance
Lightning Source LLC
Chambersburg PA
CBHW062231220526
45471CB00009B/3434